Presentation Planning and Media Relations for the Pharmaceutical Industry

Also by John Lidstone

Beyond the Pay Packet (1992), McGraw-Hill
Face the Press (1992), Nicholas Brealey
How to Recruit and Select Successful Salesmen (second edition 1983), Gower
Making Effective Presentations (Audio manual) (1985), Gower
Manual of Negotiation (1991), Gower
Marketing Planning for the Pharmaceutical Industry (second edition 1998), Gower
Negotiating Profitable Sales (1977), Gower
The Reform of the Honours System, Churchill Lecture (1998) www.churchill-society-london.org.uk/ChLect98.html
The Sales Presentation (with P. Kirby) (Audio manual) (1985), Gower
Training Salesmen on the Job (second edition 1986), Gower

Presentation Planning and Media Relations for the Pharmaceutical Industry

JOHN LIDSTONE

GOWER

Published by
Gower Publishing Limited
Gower House
Croft Road
Aldershot
Hants GU11 3HR
England

Gower Publishing Company
Suite 420
101 Cherry Street
Burlington, VT 05401-4405
USA

British Library Cataloguing in Publication Data

Lidstone, John, 1929–
Presentation planning and media relations for the pharmaceutical industry
1. Business presentation 2. Interviewing 3. Public speaking 4. Mass media and business
5. Pharmaceutical industry – Public relations
I. Title
658.4′52′024615

ISBN 0 566 08536 4

A reference block is available from the Library of Congress reference 2003054731

Typeset by MHL Production Services Ltd, Coventry
Printed and bound in Great Britain by MPG Books Ltd, Bodmin, Cornwall

Contents

List of figures

Preface

This book has been designed to help develop the knowledge and skills of people in the pharmaceutical industry in two areas of communication:

- preparing and delivering effective presentations
- handling the media.

Why this book?

It is a key part of the job of directors and senior executives in the pharmaceutical industry to make presentations – internally to colleagues, clinical staff, marketing representatives, product managers and medical and hospital teaching staff, and externally to a variety of audiences ranging from the general public to medical specialists. To make the most of such opportunities to influence these audiences, they must know how to speak well, how to present their case convincingly, and how to deal satisfactorily with their audience's questions and comments.

As well as making presentations all these people are likely to be sought out for interview by the media. Journalists and broadcasters are constantly searching for those who should have answers to important problems, so that what they say can be conveyed to readers, viewers and listeners. Moreover, all journalists regard the pharmaceutical industry as a regular punchbag and a standby for stories when there are no other stories, or during the 'silly season' when everyone goes on holiday.

Directors and senior managers likely to be invited or confronted by the media need to be equipped with the techniques and skills to handle interviews, to be able to make presentations to the media, and to be able to discuss with journalists the issues arising from them. That is part of the life of any business executive today and none more so than those involved in the manufacturing, clinical trials and marketing of life-saving medicines.

For the media bad news is good news, whether we like it or not, so it is essential for those who have to deal with or announce adverse news to be able to do so first, fast and effectively, rather than be taken by surprise. Those in responsible positions in this vital industry are expected, as a part of their job, to be able to conduct press interviews and to be able to broadcast. But if such opportunities are to be turned to advantage and not lead to disaster, they must know how to speak well and present their case and then to deal with questions and comments fluently, confidently and persuasively. Above all, what the television viewer, radio listener or reader sees, hears and reads must be believable.

This two-part book has been compiled to help you to develop your knowledge and skill in presenting your case on behalf of your industry and to exploit the possibilities offered by the media. I hope it will also provide some enjoyment.

John Lidstone

Acknowledgements

I wish to thank the following for giving me permission for the use of material:

Parker Garret & Co., '*The Past Masters*', by Harold Macmillan, pp. 58–9, Macmillan, London 1975.

Kodak Limited for publication number S-3 '*Audiovisual Projection*' and publication number S-24 '*Legibility – Artwork to Screen*'.

Carillion Communications Ltd for permission to reproduce Figure 7.1.

PART I

Preparing and Delivering Formal Presentations

Introduction to Part I

> The brain is a wonderful organ; it starts working the moment you wake in the morning and doesn't stop until you get up to make your first speech.

All of us during our passage through life find ourselves, willingly or unwillingly, on our feet in front of an audience making a speech of one sort or another. Just consider your own experiences so far. At school, for instance, you may have been cast in an end-of-term play or as head boy or head girl, had to say, 'Thank you on behalf of the whole school' to some local or national celebrity invited to present the prizes. Or at university, fancying yourself as a budding politician, you join the debating society only to discover that contributions are limited to five minutes, unaided by reference to notes!

You apply for your first job. It sounds like the opportunity of a lifetime, until you find to your horror that the recruitment process involves making a ten-minute presentation before a selection panel on a subject of their choice, followed by a question and answer session on your talk! Then, in your business, professional and social life, you find that the number of times you have to talk to small and large groups of colleagues, or to a formal audience, are legion. It may be that at the end of your induction programme, you are asked to present your thoughts and criticisms about your first three months in the company, hospital, or group practice; or you become a manager in a pharmaceutical company, so you cannot avoid addressing meetings of your staff, or talking to outside bodies which may range from groups of patients to pharmacists and even trade union gatherings. In private life the human motivations that drive you and which must be satisfied – for example, status, vanity, exhibitionism or an unquenchable desire to change the way the world is run – lead you to join a local political, charitable or professional association, union or club(s). Joining, taking part or holding office in any one of these groups exposes you to the danger or opportunity of speaking in public and either making a fool of yourself whenever you open your mouth or, by dint of advice or having a source of practical guidance, making one or a number of memorable speeches that everyone applauds and may even be quoted later.

The aim of Part I of this book is to help you to achieve this challenging but satisfying ambition. It has been designed to help you understand the communication problems you face and must plan to overcome every time you speak to a group of people; then to tell and show you, by means of text and examples, how to prepare an effective presentation. To achieve these objectives, the material has been arranged in the following sequence:

(1) The factors you should bear in mind about human communications and the way people learn
(2) The techniques and methods which, when used in the ways illustrated, enable you to prepare an effective presentation, speech or talk

(3) With examples, ways of developing your skills in actually making effective presentations.

Because visual aids, skilfully used, can give added impact to a speech or presentation, Chapter 6 is devoted entirely to visual aids; how to use them, the relative advantages and disadvantages of each, dos and don'ts and a summary of the basic rules. As one who has for more than forty years given an average of a hundred speeches and presentations in the course of any single year and in all parts of the world, I have two things to add.

First, I am still nervous in those last few minutes before I get to my feet and speak. The day I stop feeling nervous, I shall give up public speaking. Why? Because that tingle of apprehension, those last-minute butterflies in the stomach which most of us experience, are signs that we care about what our audience thinks and feels, and how it will receive what we say. Above all, shall we satisfy our audience? For if we do not do that, we shall have failed.

Secondly, speaking in public for the first time is rather like a course in physical fitness – hell at the beginning, but enjoyable once you get into practice and do it often enough. Now turn over the pages and discover the secrets of success. There are not too many, but you need to grapple them to your soul with hoops of steel.

CHAPTER **1**

Learning and Communication

The problem with human communications is that *no one thinks there is a problem!* Whether we are talking to one person or to a group, we all too often think that they not only hear what we say, but understand, agree with, and will act upon what they have heard in exactly the way we believe they should. That this rarely, if ever, happens is one of the most salutary lessons you must learn about communications.

Communication objectives

How does learning take place? The learning process depends on our communication mechanisms, so the problem of learning – whether it be mastering the technicalities of a new job or trying to fathom out what a politician stands for – is in essence a problem of communication. In simple terms, we communicate through the senses. Communication can be said to take place when *an identical message in the mind of one person is transferred to the mind of another.*

This rarely happens, for a variety of reasons that we shall explore later. Since communication is the basic ingredient affecting people's ability to learn, let us examine the objectives we seek to achieve when we communicate, the barriers that frequently prevent us from achieving these objectives and how these barriers can be overcome.

Every speaker, when considering what he or she is going to say, faces this challenge: *success depends upon my ability to communicate persuasively so that I shall achieve the objective of my talk by the end of it.*

That objective can be one of many: to amuse, as in an after-dinner speech; to persuade the electorate to vote for you in a contest which may be in your profession or in local or national politics; to induce people to change their attitudes and their behaviour because it will benefit them to do so; to get a group of people to continue doing what they have been doing in the past, for example the sales force to carry on promoting a particularly complicated medicine, or a specialist unit in an NHS hospital to take part in a clinical trial.

To achieve your purpose, you must concentrate on *five* communication objectives:

(1) to get your listeners to **HEAR** what you tell them (or see what you show them)
(2) to get your listeners to **UNDERSTAND** what they have heard or seen
(3) to get your listeners to **AGREE** with what they have heard (or even to disagree) whilst understanding what you have said or shown to them
(4) to get your listeners to **TAKE ACTION** which accords with your overall objective.

You will know if you have achieved all these objectives only if you achieve the fifth and most important objective:

(5) to get **FEEDBACK** from your listeners.

This is essential if you are to learn:

- whether they have heard what you said correctly
- how much of it they understood or misunderstood
- to what extent they agree or disagree with you and
- whether they intend to take the required action, take some other action, or do nothing at all.

Consider the most powerful feedback of all – the deathly silence that greets the after-dinner speaker's funny story when no one thinks it is!

These five communication objectives may sound easy to achieve, but everyone will have his or her own tale to add to the endless catalogue of failures. Some examples are shown in Figure 1.1.

	Message sent by communicator	Message received or reply given by receiver
1	Write down the answer to the question: 'Where are elephants found?'	An 11-plus examination candidate wrote 'Because elephants are so large, they are seldom lost.'
2	'Can I watch the Eclipse on your television?' (The speaker meant the Eclipse Stakes at Newmarket, a horse race taking place that day.)	'I can do better than that for you; here is a piece of smoked glass for you to look at it directly.'
3	A pharmaceutical company's medical representative, promoting a product for the relief of pain after bruising, said to a GP: 'This gets right into the bone.'	The GP asked (and his question was taken to be a serious one by the rep): 'How does the the patient get it out afterwards?'
4	A housewife in Chelmsford, England, asked an oil company representative to call on her because her domestic heating system had broken down. He called his technical staff and asked for a 'thief' to be brought to him so that he could take an oil sample as he suspected contamination. ('Thief' is the technical term for a device used for taking samples of liquid.)	The housewife when she heard this request replied: 'I would rather you didn't. I have been burgled once this year already and after the recent breakout from Chelmsford prison, I feel nervous.' To her, the word 'thief' had only one meaning.

Figure 1.1 Some examples of communication failure

To show you how difficult it is to transfer an identical message from the mind of one person to the mind of another, I would like you to participate in the following test.

Its objective is to test how well you:

(1) *hear* or *read* correctly what you hear or read
(2) *understand* correctly what you have heard or read
(3) *agree* or *disagree* with what you have heard or read
(4) *act* or *react* to what you have heard or read.

Take a plain piece of paper and on it draw four squares as follows, numbering them 1, 2,

1	2
3	4

3, 4 as shown below:
Read out each of the following four questions or, better still, ask some one else to read them out to you. After you have read or heard each of the four questions, write down your immediate response to each one in the appropriate square.

Question 1: 'You are the captain of a ship sailing due north in mid-Atlantic at a speed of 12 knots. After steaming at this speed and in this direction for 30 minutes, the captain gives the order to the engine room to alter course through 180 degrees and then to maintain the same speed on the new course for one hour. After another hour, the captain orders the engine room to change course through 180 degrees back on to the ship's original course of due north.'

Now, in the square numbered 1, write down either: the age of the captain in years or 'I don't know'.

Question 2: In the centre of the square numbered 2, draw a horizontal line thus:

____

Having done that, write down the first and last letters of your own surname at each end of the line you have drawn.

Question 3: In the square numbered 3, write down the figure **1** followed by the plus sign +, then another **1** and then an equals sign = as follows: **1 + 1 =**

Then write down what you believe is the answer.

Question 4: Now for the final question, to be answered in the square numbered 4, write down your immediate reaction to the phrase: 'Paris in the spring'.

Question 1

In answer to the question about the captain of the ship most people will write down: 'I don't know'. Most English people, that is. On the other hand most continental Europeans will answer differently. They will write down their own age. The reason for this can be found in the contrasting methods by which English and other Europeans are taught. The English are taught by what are known as Aristotelian principles. Based on the writings of the Greek philosopher, Aristotle, this method involves giving the students all the background to a subject or arithmetical problem *first*. The answer or solution comes at the end of a teaching session. Remember what happened in your own school days? Now one of the consequences of this is that English people tend *not* to listen to the beginnings of sentences or the beginning of a speech or lecture and, most irritatingly, they very frequently do not listen to and register correctly the name of a stranger to whom they have been introduced for the first time. After someone has been introduced to an English person, the latter will be heard to say; 'What did he say his name was?' Now this is probably what happened to you if, in answer to the question 'What is the age of the captain?' you wrote down 'I don't know'.

Otherwise you would have realized that 'You are the captain of a ship' means YOU, the person listening to or reading, so the answer is obviously your own age. Unlike the English, most continental Europeans are taught by the scientific method, which is the opposite of the Aristotelian. The scientific method starts by giving students the essential information and answers to the problem *first*, and thereafter the bulk of the teaching is devoted to showing and explaining how that solution was arrived at. Consequently most European students, and in later life as adults, listen attentively to a speaker from the very beginning. Otherwise, as they know from their schooldays, what follows will be neither logical nor sequential. This illustrates dramatically that people *do not hear correctly* what they have just been told. They do not even listen to or take in the first words you say to them, or what they have read at the start of a book, article or paper.

Question 2

How did you reply to this question? The majority will put the first letter of their surname at one end of the line they have drawn and the last letter at the other end. My surname LIDSTONE would be represented thus:

L________________E

However, this is not what you were actually asked to do which in my case would look like this:

LE________________LE

Why do we fail to get this right? Quite simple really. What I asked you to do did not make sense or sound logical. So your mind translated what you heard or read into what did make sense and you put down the result, *rejecting* what I actually asked you to do.

The communication lesson here is an important one. People will *understand* if you convey to them your ideas in a way that makes sense to them. It is logical for the first letter of your name to come at the beginning and the last letter at the end, isn't it?

Most misunderstandings that occur in human communications are due to our failure as communicators to put ourselves in the position of the listener and imagine how he, she or they will understand what we are about to say or have just said.

Question 3

What was your answer to this question, $\mathbf{1} + \mathbf{1} = ?$

Most people, recognizing this as a straightforward arithmetical problem, write down

1+1 = 2. Is that what you did? I expect so, like the majority of us. Yet there is at least one other alternative answer and possibly a number of others. Among an audience I talked to about public speaking were three art students. When they saw 1 + 1 = it appeared to them as an art form and, without hesitation, they wrote down **1 + 1 = six straight lines**

What is the communication lesson to be learned from this reply that may surprise you, yet is just as correct as 1 + 1 = 2? Simply this. Before communicating with another individual or speaking to a group of people, we should *never make assumptions* about what they should think, or assume that a statement is so obvious that everyone will agree with it. It is much better, always, to put ourselves in the position of our listeners and ask: 'How will he, she or they receive and perceive what I am about to say or show?'

Question 4

What did you write down in response to the phrase 'Paris in the spring'? This evocative phrase has produced thousands of different replies in all parts of the world, ranging from the predictable, 'flowers', 'the Seine', and 'love' to, on one occasion, 'sausages'. This surprising reaction came from a member of the marketing team of a well-known meat company which had asked him to take part in a television commercial to be made in Paris. This therefore was his most vivid, most recent and most powerful memory of that city! Incidentally, never have I been given the prosaic reply: 'The capital city of France at a particular time of the year'.

This highlights another communication lesson. We should always weigh our words and phrases carefully in case they evoke quite different responses from the one we intend. These four communication exercises demonstrate that our four objectives when we make a speech or presentation, to get people to:

HEAR what we say (or to see what we show them)
UNDERSTAND what we mean
AGREE with what they have heard
TAKE ACTION in accordance with our overall objectives

are not as easy to achieve as perhaps we thought. Let us now look a little more closely at these communication barriers, what we can do about them and how we can make our presentations more effective by planning them from the listener's point of view.

The main problems in communication

The first problem is to recognize that, despite what many people believe, communicating successfully is not easy. The second problem is to accept that the onus is on the communicator, not the receiver to achieve successful communication. A number of specific difficulties arise that may prevent the achievement of each objective:

Objectives	*Difficulties*
HEAR (or see)	People cannot concentrate for long periods on the spoken or written word. People pay less attention to what appears to them unimportant.

UNDERSTAND	People often make assumptions based on their past experience. People often do not understand the speaker's jargon. People misunderstand more easily when they hear but do not see. People often draw conclusions before the speaker has finished.
AGREE	People are often suspicious of others with an interest in selling something. People do not like being proved wrong.
ACT	People do not easily change their habits. People fear the results of taking wrong action. Many people dislike making decisions.
FEEDBACK	Some people deliberately hide their reactions and what they really think. Appearances can be deceptive – a nod may not always indicate agreement or understanding and can mask ignorance or indecision.

These difficulties are common to both the communicator and the listener. Neither we who communicate, nor our listeners:

- like to be proved wrong
- pay attention to what is unimportant
- change our habits easily
- understand other people's jargon.

If we examine the communication process we can better understand how it works, how failures in communication arise and what we can do to be more effective and successful as communicators.

Figure 1.2 illustrates the way we communicate. Messages are received through our senses of *sound, sight, feel, smell and taste.* We then form impressions and assimilate or associate them with other information and ideas stored in the brain. Before we respond to what has been communicated, the brain reacts in a specific sequence to this new information. It scans memories of past experiences and finds the frame of reference or memory that relates most closely to the new information. The new information is then sent to join the memory bank or frame of reference chosen.

If it is associated with what the memory perceived, the new information is analyzed and subsequently fitted into the existing memory pattern. As a result of this filing system of the brain, the existing memory may:

- be strengthened
- change for the better
- change for the worse.

Examples of this memory bank at work are legion. The politician whose party is seeking office, after five years in opposition, paints word pictures of carefully chosen unpleasant features of life which their listeners will not only recognize but about which (the politician hopes) they will agree with them. The chief executive talking to his or her assembled managers identifies with them as he or she describes some of the mischief he or she got up to as a young manager.

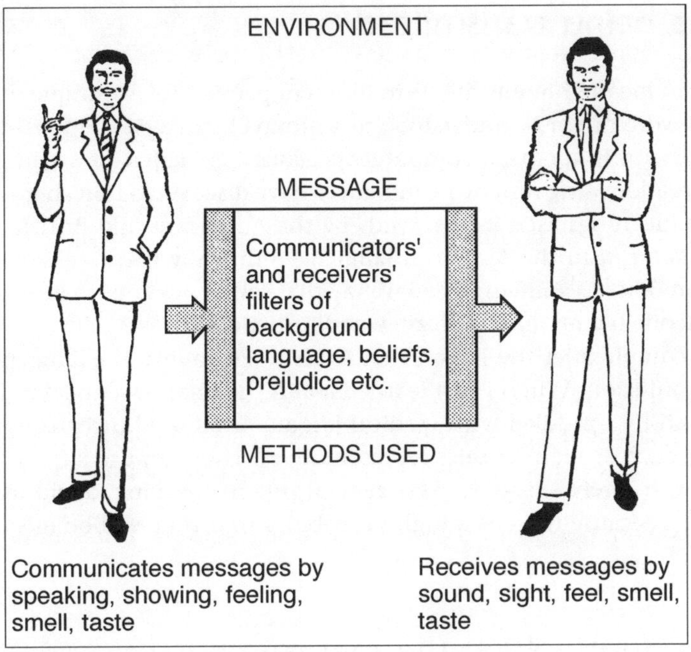

Figure 1.2 How the human communication system operates

Apart from reinforcing what we believe, other factors influence the quality of our communications. There are five main elements that, as we have seen from Figure 1.2, can lead to failures of communication:

(1) the values of communicator and listener
(2) the message being transmitted
(3) the filters through which the message passes
(4) the methods of communication used
(5) the environment in which the communication takes place.

The importance of values

Our background, education, beliefs, ethical standards and prejudices all affect the way we communicate with one another. Thus two people in an audience looking at the same object or picture, or listening to the same story, may perceive and react to it in quite different ways. Because of this, it is vital for a speaker to try to perceive whatever he or she wants to say or show through the eyes and minds of the people who will comprise the audience. It is a knowledge of the audience's ideas and experiences that enables a speaker to communicate successfully. This point serves to emphasize the importance of doing your homework as to the nature of your audience in advance. Men or women, or both? From diverse backgrounds? If they are called doctors, are they medical people, or from other disciplines as is often the case with a continental European audience?

The message being transmitted

The same words can mean different things to different people. Take a simple noun like 'run'. Imagine how that word might be understood by a group of bankers, as opposed to a group of theatre producers – not to mention computer specialists, cricketers or angling enthusiasts!

Many speakers addressing American audiences have discovered that there are important differences in the meanings ascribed to words by the Americans and British. For example, there is no word *fortnight* in the American language – they say two weeks; we speak about holidays, Americans about *vacations*; when we say that someone is out to lunch, they are – to an American someone *out to lunch* is crazy, weird, intoxicated!

Add to these complexities the jargon that tends to creep into the language of business and the result is confusion. When I first learned about the pharmaceutical industry it was – for that matter it still is – riddled with medical jargon, often used incorrectly by those not medically trained, such as medical sales representatives. They can use jargon quite wrongly to impress doctors. It rarely succeeds. As a general rule in speaking, avoid using technical words – or jargon – because there is a high probability that they will be misunderstood by your listeners.

The filters through which the message passes

All of us tend to think more about ourselves than about the person or group we are communicating with, and our choice of words, how we express them and the meaning we impart to them reflect this. Yet our words, beliefs, prejudices, and jargon can set up filters that corrupt the message sent and received. One example of this is the word 'marketing', used by an ever growing number of people to mean – what? To some it is a word that represents the essence of business, that is the identification and satisfaction of a customer's needs at a profit. Yet I know of at least one well-known computer manufacturer who mistakenly uses it to describe their sales managers, whom they call 'marketing managers'.

I was once invited to give the annual 'marketing address' to a group of these managers. I prepared my talk on the reasonable assumption that they were what their title suggested. I started to speak to them about marketing issues such as SWOT analyses, perceptual mapping and so on, only to find that such terms were a closed book to them. They had no idea what I was talking about!

Methods of communication

No two individuals hear, see and feel with equal efficiency. You can tell some people something and they understand immediately. Others have to be told, shown and then asked to play back their understanding of what they have heard and seen before a message gets through. For this reason, when communicating ideas, a speaker making a presentation should always involve at least two of the senses through which listeners receive a message. As a speaker you can express your ideas not only verbally. Some of them, including those you particularly want your audience to remember, will become more firmly fixed in the minds of the audience if you distribute them as a handout or show them as a chart, or on an overhead transparency or slide.

The environment

The environment in which a presentation takes place can have a profound effect on its outcome. If you are addressing members of a company and different levels of management are present, what you say is probably going to be digested at least *twice* by the junior managers. First they will hear what you say; then they will try to guess how it will be received by the senior managers present. Likewise, if as a medical director, product manager or medical representative, you are giving a presentation to a group of GPs in a group practice, the chances are high that some of audience will come in and go out. They are at the beck and call of patients or emergencies. In such situations, make sure that you have handouts, or even an audio-tape version of your presentation available for those who may have missed parts of it.

When a very large audience has congregated to hear you speak, it is likely to react almost as one person. Indeed some speakers frame their presentation from the view of one person. This technique can be an effective one so long as your research into the nature of your audience and what they expect has been thorough.

How can we communicate more effectively?

The communicator's role is that of a teacher imparting his or her point of view to the listener. This process of helping people to learn can be made much easier by understanding and using the laws that govern learning, which are the laws of: *effect, forward association, belonging* and *repetition.*

EFFECT

A listener will more readily and willingly learn if your message shows how to satisfy an acknowledged *need*. US President Franklin D. Roosevelt's 'fireside chats' are an example of how to do this. When campaigning for election in the 1930s, Roosevelt used to speak with tremendous effect on the radio to his unseen listeners. He never started by saying such things as 'I am standing on a programme of this, this and this, so vote for me'. No. He would begin by identifying the needs of average Americans and what they wanted. 'You good people of America, you need jobs to provide money to keep your family fed and clothed and to uphold your self-respect, you want.'

I have been told by Americans who listened to these fireside chats that listeners not only nodded as Roosevelt spelt out their needs, but started saying, 'Yes, yes, that's just what I want'.

FORWARD ASSOCIATION

People tend to remember things in the order in which they first learned them, especially if they are arranged in a logical sequence, for example:

MY NEEDS ⟶ THE BENEFITS WHICH SATISFY THEM ⟶ THE TECHNICAL FEATURES WHICH PROVIDE THESE BENEFITS

The reason so many speakers fail to engage their listeners from the start is that, unlike President Roosevelt, they talk about their own needs and what they want from the audience, not their listeners' needs. Or they start describing technical features that are uninteresting from the listeners' point of view. If a GP tells one of his patients that 'For the benefit of this surgery, you must telephone for an appointment at least ten days ahead', the patient's attitude is likely to be 'To hell with what you want, what about *me*, for heaven's sake'.

BELONGING

People learn more quickly and easily what relates to their own experiences, for example 'an air conditioner is like a refrigerator element with a fan to circulate the cooled air' or 'The plans I am about to describe are like those you make for your annual holidays to ensure that you will enjoy them'.

REPETITION

Contrary to what is often thought, constant repetition of a statement or a warning does not result in people learning, as pronouncements about dangerous driving, excessive drinking of alcohol, smoking, eating food which can lead to obesity and ill health, and so on, testify. Repetition is only useful as a means of getting people to learn if it is used in conjunction with one or more of the three laws of learning already described. A speaker using repetition plus another law might say: 'So let me repeat what I said at the beginning of this talk: if you are serious in your wish reduce the dangers of obesity for your children, then here is what we all have to do.'

Research shows how people forget what they learn:

- 38 per cent of what they hear within 2 days
- 65 per cent of what they hear within 8 days
- 75 per cent of what they hear within 30 days.

But they tend to remember what is important or of special interest to them. We can help them remember by:

- ensuring that our first and last impressions upon them are both favourable and positive
- starting a meeting, if possible, by summarizing progress made at earlier meetings
- giving them a general idea of a proposition before moving on to points of detail
- involving them from the start by:
  - talking about 'your problems' and 'your requirements' rather than 'what I want'
  - obtaining feedback, so that we know how well we are communicating, and thus can judge whether it is necessary to rephrase our remarks or repeat what was said earlier
  - using more than one sense (for example speech and visual aids and sometimes – when it concerns food or taking a medicine, for instance – taste)
  - planning to communicate.

Points to remember

(1) Unless someone *hears* what you say, there is no communication.
(2) You do not communicate just words, the whole person comes with them.
(3) Talk to people in terms of their own experiences and they will listen to you.
(4) When you have difficulties in getting through to people, it is a sign that your own thinking is confused, not theirs.
(5) When you fail to communicate, it is not your words that need attention; it is the thoughts behind them.
(6) Know what your listeners expect to hear and see before you start talking.
(7) Your communication is always more powerful if it appeals to the *values* and *aspirations* of your listeners.
(8) If what you say conflicts with the beliefs, the aspirations, the motivations of your audience, it is unlikely to be received or listened to at all.
(9) It is not what is written on the pages of your presentation that matters – it is the proportion of it that comes off those pages and enters the listener's mind and stays there.

The listeners' point of view

One of the effects of the growth in communications, particularly those involving the visual sense, is that we have become accustomed to certain standards of performance from those who address us. We may not agree with what is being said, we may not even be interested, but we cannot fail to notice the style used by the speaker. We remember the image long after we have forgotten the content.

As a result it is impossible to say nowadays that any form of presentation will do. Rightly or wrongly, an audience, whatever its composition, will judge a speaker's ability, and that of his or her company or organization, by the kind of performance they deliver. This is not to say that the substance of the speaker's proposals is unimportant, rather that the impact is disproportionately enhanced or diminished by the quality of the presentation. You are in a sense only as good as the ideas for which you gain acceptance. There are four categories of speaker:

(1) Those who do not bother about what they are going to say or how they are going to say it. They exhibit a 'couldn't care less about the audience' attitude.
(2) Those who 'put on a show' but convey very little.
(3) Those whose material is good but badly presented.
(4) Those who have something to say and present it well.

The judge of a presentation is the audience. No two audiences are the same. The individuals in an audience will differ in attitude, but whatever their personalities and job responsibilities may be, they all react to presentations. There are certain mental demands that have to be met before they will give their acceptance. In addition, they are affected by what they see, what they hear and how they feel. All these can be summarized as the listeners' point of view, and its basic elements form a sequence that can help a speaker to structure his or her presentation.

THINKING SEQUENCE

The thinking sequence that the listener's mind follows consists of seven steps. A good presentation takes account of them all.

1 *I am important and want to be respected*

Every member of the audience wants the respect of the speaker. Without it, the speaker is lost.

2 *Consider my needs*

Any proposal is judged by the listener in terms of his or her own priorities and values. These are determined by what they want to achieve: (a) in their work (b) as a person.

Consider a typical group medical meeting.

Your company sells a leading product for diabetes. You are the medical information executive working in the company's medical department and attending this meeting to give technical support to the Product Manager responsible for your diabetes prescription medicines. You will deal with any medical or technical questions.

The meeting is being held in a group practice in a suburb of Manchester. Your audience will consist of three GPs, with two more hoping to attend if circumstances permit, plus the group practice manager and the group practice nurse. As it turns out the two GPs, who hoped to come, do put in brief appearances, but are both called away to deal with emergency calls. As the Product Manager begins his introduction, here are the thoughts going through the minds of two of the three GPs present.

First GP's immediate work needs:

> I hope he comes to the point quickly. I've got twelve patients to see this morning, three home visits after lunch and a BMA meeting at seven this evening.

Personal needs:

> Oh damn, I promised to help Harry revise his biology for tomorrow's A level exam. What is this woman talking about now?

Second GP's immediate work needs:

> Why did I ever agree to come to this meeting? There are two medical reps down to see me already today. Anyway, there's an article in *The Lancet* about mature diabetes. It would probably have been better to read that than be sitting here listening to this stuff. I expect I've heard it all before.

Second GP's personal needs:

> What did my wife say to me last thing this morning? Blast. Now I remember, she wants me to be sure to be home before six o'clock so that she can visit her mother. And I promised to pop in and have a drink and discuss a patient with Clive Boyd [a cardiac specialist].

A presentation will make little impact if the listener cannot see that it has regard to improving his or her lot. In a business context their needs will relate to such things as improved profitability, higher sales, lower costs, better staff relations and so on. For a hospital they can range from speedier handling of accident and emergency patients to shorter waiting lists for consultants and greater availability of beds. The listener wants to know early in a presentation that it will deal with the issues uppermost in their mind. If so, they will give the speaker their willing attention. In the same way their final decision will depend on their answer to the question: 'Will my needs be met by this speaker and their proposition?'

3 *Will your ideas help me?*

If the listeners' attention and interest have been engaged, they are keen to know how the speaker's proposals will help them achieve the results they seek. They want to know what the speaker's proposals will do for them and their practice, their hospital, their pharmacy, their organization.

4 *What are the facts?*

The fourth step in the listener's thinking process arises from the previous one. They want to know how the speaker proposes to ensure that the promised results are forthcoming. They may demand evidence that these results have been achieved in cases similar to their own. They also want to know what is required from them – action to be taken, time commitment and so on.

5 *What are the snags?*

It is an integral part of the listeners' decision-making process to consider the possible disadvantages, or consequences, arising from the speaker's proposals. If any come to mind that they cannot imagine being overcome, they will often express them in the form of objections – in a group, though, there is a bigger chance that such objections will remain unvoiced. Few people want to risk offering an opinion that may be unpopular.

6 *What shall I do?*

If all the previous steps have been covered satisfactorily, the listener now faces a decision: 'Do I accept or reject these proposals?' In making the choice, they will have in mind their needs in respect of their job or their personal life and decide accordingly. Where there are several sets of proposals to consider, they will prefer those that, in their eyes, best meet their needs.

7 *I approve*

If steps 1 to 6 have been competently handled from the listeners' point of view, they will decide in the speaker's favour.

THE IMPORTANCE OF THE LISTENERS' POINT OF VIEW

The seven steps mentioned above represent the path that the human mind follows before it will give an approval. The problem facing speakers, however, is that it is difficult to present proposals in that sequence and with that kind of emphasis. It is natural to concentrate on the ideas being promoted, while of course the audience is more interested in what they want to achieve themselves and what the speaker has to say about it. Consequently the audience loses interest very quickly, their attention wanders, and they end up rejecting both the proposals and the speaker.

Structuring the presentation around the listeners' point of view can go a long way towards gaining the audience's attention and interest, persuading them of the value of the speaker's proposals, meeting their objections, and leading them to the desired conclusion. Remember, though, that adherence to these seven points is not a natural process; we are all far more concerned about ourselves and our needs than those of others. So before every speech, five-minute talk or, especially, lengthy presentation, remind yourself of these seven points.

OTHER CONSIDERATIONS

When presented with proposals the human mind not only thinks along certain lines, it is affected by what it sees and hears and, to a lesser extent, by sensations of touch, taste and smell. In formal presentations sight and hearing are of the most concern to a listener.

Sight

Listeners react to their first sight of you as a speaker. They expect your dress, facial expressions and gestures to match their mood and the content of your presentation. They look for signs of confidence. Consider the position from the listeners' point of view. They see:

(1) *How you are dressed.* Are you wearing bizarre clothes which will distract your audience from what you say to them? Try to dress neatly and in the same way as the group you will be with. Richard Branson's casual dress style may have been acceptable to those who worked for him, it has not spread. And whereas at one time business people 'dressed down' on Fridays, the trend now is in the opposite direction. People want to feel comfortable in the presence of their peers and, much more importantly, when visited by their customers!

(2) *Your mannerisms.* Always be yourself but avoid distracting mannerisms. Some speakers wave their arms up and down when talking, like a merchant in a bazaar. Gestures should be few and powerful, and used to emphasize specific points.

Listeners find it much easier to concentrate on things that they can see. But what they see must be *understandable, simple* and *professionally handled*. They have greater confidence in a speaker who looks at them. Keep in touch with your audience by maintaining eye contact. Nothing loses an audience so quickly as the sight of a speaker whose head is buried in his notes all the time.

Hearing

There are two important differences between a public presentation and a normal conversation. In a conversation you can ask your listener if they understand what you have just said and they can ask you to repeat something if they did not hear you or understand. But in a public presentation this is not always possible. You can never rely on a member of the audience to interrupt and ask you to 'explain that in simple terms' or even to shout out 'speak up, we can't hear at the back'. So you have to make sure that your message is heard and understood *first time*. To that end, remember the following:

(1) Speak louder than you would during a normal conversation. Adapt the scale of your presentation to the size of the room or hall and to the size of the your audience. If you rehearse in the room where you are going to speak, ask a colleague to sit at the back and check that he or she can hear you clearly from there. But bear in mind that when the room is full of people you will need to project your voice even more strongly.
(2) Articulate distinctly and emphasize the last words in each sentence. Inexperienced speakers tend to drop their voice at the end of a sentence. If it contains the most important part of your message and no one hears you your presentation has failed.
(3) Audiences expect you to use language they can understand. Keep it simple always.
(4) Don't speak too fast.
(5) Vary the pace and the pitch of your voice to maintain people's interest.
(6) Use pauses. In public speaking there is no device more effective than a pause. It gives the audience time to digest what you have just said, and especially what you have shown, and it gives you time to glance at your notes. It keeps the audience expectant about what you will say next. Incidentally, the speaker usually perceives a pause as lasting much longer than the audience does.
(7) People dislike having to listen to a presentation that is read.

Success in all formal presentations and speeches is founded upon understanding the listener, thinking about their point of view and planning what you will say and show with the aim of meeting it. Good ideas, however sound or well argued, will not succeed on their own merits alone. They have to be presented *attractively, clearly* and above all *persuasively*. This means combining a listener-based structure with presentational skills to achieve your planned objective.

In his book *The Past Masters* Harold Macmillan gave a salutary description of his metamorphosis as an orator. Here is an excerpt from the chapter he wrote about his mentor, another British Prime Minister, David Lloyd George.

> In 1924 he [Lloyd George] still held the rapt attention of Members [of the House of Commons], who flocked in when his name came up on the 'ticker' in the various rooms.
>
> I first got to know him in a peculiar way. In my rather hesitating, shy and stilted style I delivered a speech (not my maiden speech, but in the early years of the Parliament) to which by chance he happened to listen.
>
> As usual the House was very thin for new Members, especially Members on the Government side who were only called at times when the audience was small. Into my speech I put all the thought and care that I commanded. It was on an economic subject, dealing with unemployment and its possible cures. I had worked at it and it was received with respect. Later that evening Lloyd George came up to me and said, 'Macmillan, that was an interesting speech of yours'. I was naturally flattered. He continued: 'If you don't mind me saying so, you have no idea how to make a speech'. I answered, 'Will you tell me', and with great kindness he took me up to his room to

give me his fatherly advice. It ran something like this. 'First of all you are a new Member. You always speak in a thin House, probably in the dinner-hour. Even I am not called till six. Never say more than one thing. Yours was an essay, a good essay, but with a large number of separate points. Just say one thing; when you are a Minister two things; and when you are Prime Minister winding up a debate, perhaps three. Remember your own position. There will be few listeners. What you want is that somebody will go into the Smoking Room and say, 'You know, Macmillan made a very good speech'. 'What did he say?' someone will ask. It must be easy to give a ready answer – one point. Of course you wrap it up in different ways. You say it over and over again with different emphasis and different illustrations. You say it forcefully, regretfully, even perhaps threateningly; but it is a single clear point. That begins to make your reputation'.

Then he went on to explain that my speech had been delivered in a monotonous way without light or shade. 'What is the right way?' I asked. 'Why, there must be continual variation; slow solemn phrases, quick, witty amusing passages'.

Above all say to yourself as you get up, 'Vary the pace and vary the pitch'. This is the heart of the whole matter. Finally, don't forget the value of the pause.'

I have never forgotten his advice and watched him over and over again with admiration. He would start generally in a low tone, and knowing that he had to give time for Members to come into the Chamber from the Dining Room or Smoking Room, or the Library, his opening sentences were of a vague and introductory character (just as in the opening few minutes of a West End play). I remember his explaining this to me in some detail. 'You can't have the murder in the first few minutes because the stalls and so on are still coming in. The butler comes on; the telephone rings; the soubrette comes on; and after five or ten minutes when the audience are settled down the play begins.'

When I protested 'But how do you do this?' he replied, 'Oh it's quite easy. The speaker who has just sat down in a thin House is probably a serious but not exhilarating speaker. You say that you have listened to his speech with great interest and sympathy. There is just one point you didn't understand. Why did he say that there is no difference between black and white? He rises to explain. You courteously give way. When he sits down, if the House has filled up, you accept his explanation with a suitable apology. If not, you say that of course you now understand, but still, why did he say that there is no difference between green and yellow? He rises, a little angrily. You give way. The little comedy is repeated. Then, your audience having arrived, you start your speech.'

CHAPTER 2 Preparing your Presentation

You always start with an advantage if you have been specially invited to speak to an audience. But the relationship is reversed if it is you who apply to make a presentation. This is often the case when you, or your company, want to influence a particular body of people. Such presentations are really selling opportunities. Your relationship with the audience becomes an unnatural one because:

(1) You have sought out your listeners, who may be GPs attending a meeting to which they have been invited by a pharmaceutical company representative or director, or you may have asked to speak to specialists at a hospital.
(2) You want them to take action(s) favourable to you or to those whose interests you represent, you company, your profession, your department.
(3) You may have to replace their ideas with your own, and some of those held by your audience may be entrenched ones.
(4) Because you yourself have asked to present your case, you can often feel as though you are on your own with little or no support from those listening to you. Indeed, sometimes they may be hostile.

These problems, which you share with everyone who seeks to sell products (like the medical sales force), services or ideas, create *tension*, and often make you act out of character. You start talking too fast or too loudly; you look down at your notes instead of maintaining eye contact with your audience; you end up concentrating too much on your ideas rather than the needs of your audience.

Planning helps to reduce tension and ensures that your presentation will be focused on what your audience 'requires' or 'needs' and wants to hear rather than on what you, the speaker, want or require.

What should you prepare?

Since planning is simply a process of thinking about what you are going to do and how you are going to do it *before doing it*, the first thing to plan is the *presentation objective*. Although the ultimate purpose of every presentation is likely to be to *persuade* and to *influence*, you need to define your objective in your own mind in precise terms. A useful way of doing this is to ask yourself the question: 'At the end of my presentation, what do I want my audience to do?' The answer provides an introduction to your presentation, keeps you to the point in the middle and, finally, supplies you with the inspiration for your closing words. Now you can write down two, or at most, three things your objective could be expressed as. Here are some examples:

'At the end of my presentation, I want the audience to have sufficient information about the research findings to decide in favour of undertaking clinical trials by a specific date.'

'At the end of my talk, I want the audience to agree that a change in prescribing for patients with osteoporosis will reduce dosage and improve their way of life significantly.'

'At the end of my presentation, I want the audience to support our campaign to raise finance for a new ward at Stoke Mandeville Hospital.'

Studying your audience

Having defined your objective, ask yourself these questions about your audience so that you can decide what you need to know:

(1) Who will comprise the audience? – all men, all women, mixed?
- age group?
- what do they do?
- how many ?
- What are their backgrounds, culture, specialisms?

(2) What do they think?
(3) What do they think they know?
(4) What do they really know?
(5) What do they expect from me?
(6) Will they want to ask questions during or after my presentation?
(7) If they ask questions
- who is likely to be an ally?
- who is likely to be hostile?
- who is likely to be neutral or indifferent?

(8) How can I influence them?

Title

A title provides a point of interest for the audience and a focus for what you want to say. And the organization responsible for staging your talk may want to send out an advance notice and for this it will need some details. Titles for talks can be provocative statements or open-ended questions. Some examples are:

'The medical sales force – does it have a future?'
'Are some of the old cures better and need bringing back?'

Structure

Any presentation should be based on the listener's point of view discussed in Chapter 1. It should comprise three main parts: the *beginning*, the *middle*, and the *end*. Figure 2.1 illustrates their content and function.

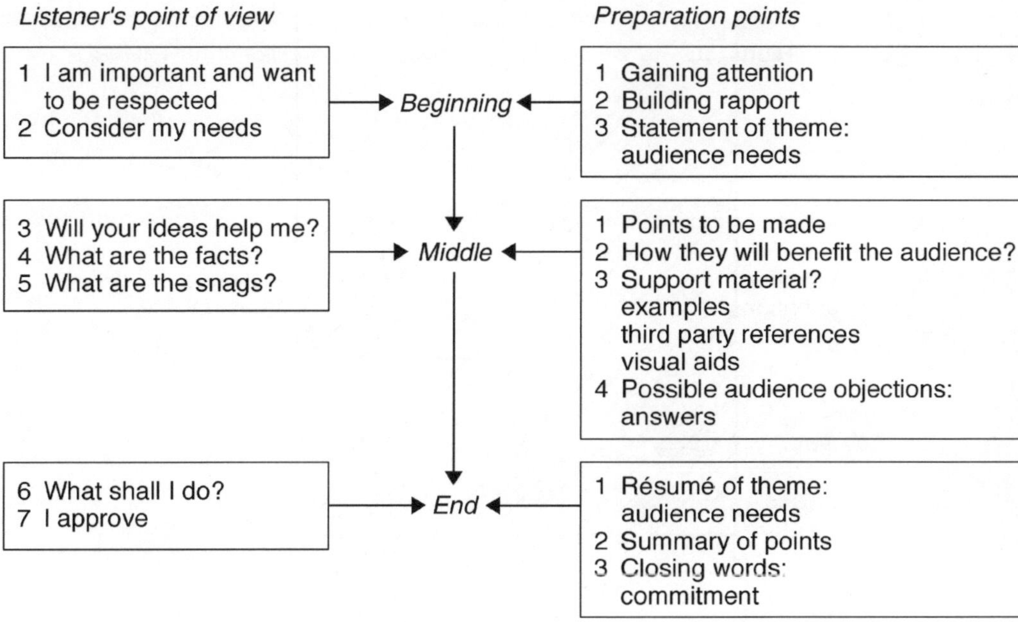

Figure 2.1 Structuring a presentation

How does a structure help? Apart from reducing tension and ensuring an audience-oriented presentation, structure has other important advantages for a speaker:

(1) It enables the audience to follow the presentation easily, being based on an initial outline of the theme, followed by development of the theme and concluded by a summary of the theme and the key points made, with a request for action.
(2) It ensures that all the audience's psychological demands are met.
(3) It provides a secure framework to fall back on if any questions from the audience during the presentation should lead the speaker astray.
(4) It provides a disciplined and logical basis on which the speaker can plan the presentation.

The next three chapters explain how to begin, develop your presentation in the middle and how to end on a high note. A presentation divides almost naturally into three parts and many speakers prepare their talks on the basis of the three-times-three (3 × 3) rule:

(1) Select up to three, but never more than three, main points or objectives for your talk.
(2) Tell your listeners three times, using the famous speaker's formula:
    - Tell 'em what you're going to tell 'em (at the beginning)
    - Tell 'em (in the middle)
    - Then tell 'em what you told 'em (at the end).

This formula acts as a double insurance. When an audience is listening to a presentation, it goes through a known pattern of attention to what is being said. At the beginning, audience attention is *high* but after ten minutes it *diminishes*. Then people tune in again and listen closely to the speaker's final remarks. This attention curve is shown in Figure 2.2. Armed

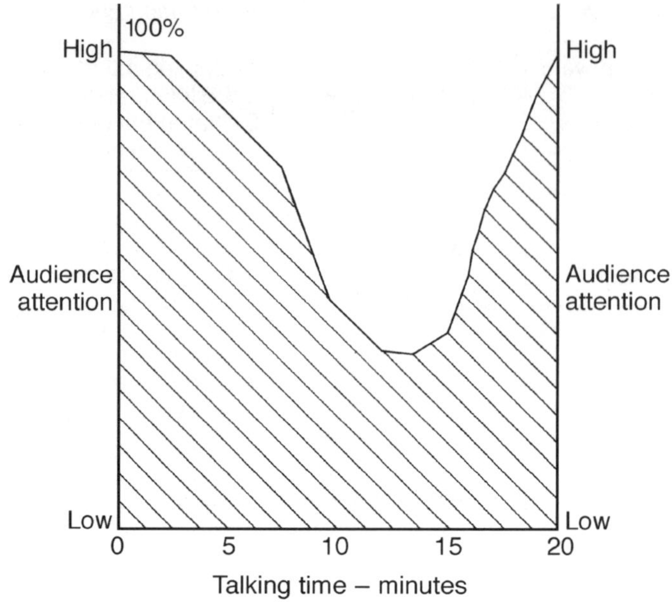

Figure 2.2 Audience attention curve

with this knowledge, you should plan to cover the important points at the start and re-state them at the end.

Preparing your notes

First of all, dismiss from your mind the idea that you can rise to your feet and make a speech or give a presentation extempore. That way lies disaster. One of the secrets of successful speaking is knowing how to write down what you propose to say in notes to which you can refer easily. What you don't want is to become the prisoner of a sheaf of closely written material to which your eyes are glued, thus failing to hold the attention of your audience.

But it is equally dangerous to try to memorize your speech so that you can dispense with notes altogether. Never, never do that. The result will be either catastrophe or a presentation that lacks warmth, sincerity, emotion or spontaneity. Describing one speaker who did this, the late Sir John Simon remarked: 'He had everything except one thing. Nobody believed he believed'.

For a talk or presentation of any length, write down what you plan to say in detail. You will probably start doing this from the moment you agree to the speaking engagement. It is not a bad rule to assemble a hundred thoughts or ideas and then discard ninety of them. Select the key words or sentences that summarize each section of what you want to cover. As you do this, speak them aloud to yourself to hear how they sound. Some words and phrases that read well do not have the same impact when spoken.

Now print these key sentences and words in large capital letters on cards so that when they rest on the table or lectern from which you are to speak, they will be easy for you to glance down at and read. Use either blank postcards or lined index cards sized 6 × 4 inch (100 × 150mm). Keep the number of words on each card to a minimum. Make sure, though, that you write out in full your *opening sentences* and *closing words*. These are the two most important parts of your presentation; try to memorize them so that when you start and finish you can look at your audience and have full eye contact with it.

Items that should never be abbreviated in a speech or presentation are quotations and statistics – *get them right*. If you decide to use a well-known quotation from, say, the Bible, Shakespeare or a famous author, never rely on your memory because it is sure to let you down and you will misquote. And, if you do, there are always people in your audience who will know that particular quotation. If they find you getting one thing wrong, they are likely to dismiss everything else you say as inaccurate.

The same applies to statistics. Always be accurate with any information you quote. Because figures on their own are difficult to grasp or interpret, try to relate them to images that your audience can easily visualize. Instead of saying 'about 22 yards' refer to 'about the length of a cricket pitch'; instead of '580 feet', talk about 'the height of the BT Tower'.

It is always tempting when preparing a presentation to try to be as up-to-date and as topical as possible by referring to the evening's newspaper headlines or a story that has broken on the latest television news bulletin. Do not quote news items if they are less than twenty-four hours old – surprisingly few people will have had the same motivation or interest to read them as you have. If you do use such stories, you must give the whole news item so that those who did not hear or see it can understand it as fully as you do.

Assembling your notes

Once you have collected your thoughts and ideas and printed them on cards to create the final material for your presentation, you should assemble them in such a way that they cannot get out of order.

(1) Number each card in sequence, and on the right-hand side indicate by number any visual aid you plan to show at that stage of your presentation.
(2) Punch a hole in the top left-hand corner of each card and then link them all together with a treasury tag (see Figure 2.3).
(3) Review your whole presentation card by card. You may wish to highlight certain words, passages, or figures by underlining them or by using highlighting colours.

Finally, never, never delegate the task of preparing and assembling your notes to anyone else, even your secretary. However good they may be, it is you who will be making the presentation, not them. A former colleague from my Shell days, the late Sir John Greenborough, told a story against himself which provides an awful warning to anyone tempted to delegate this task.

He had been invited to speak at a City of London livery company's annual dinner. He dictated to his secretary the speech he intended to give and asked her to make him an original and two copies of it. He collected the three copies later and went off to the dinner. The next morning the secretary asked him how his speech had been received. 'Well', replied Sir John, 'the first twenty minutes, very good, during the second the audience became a little fidgety and towards the end of the one hour I spoke I sensed pronounced signs of open hostility. Strange. I thought it was quite a good speech. I wonder what could have gone wrong. Have you any ideas?' To which his secretary replied, 'Well, Sir John, I did give you the original and two copies, as you asked, for your twenty-minute speech!'

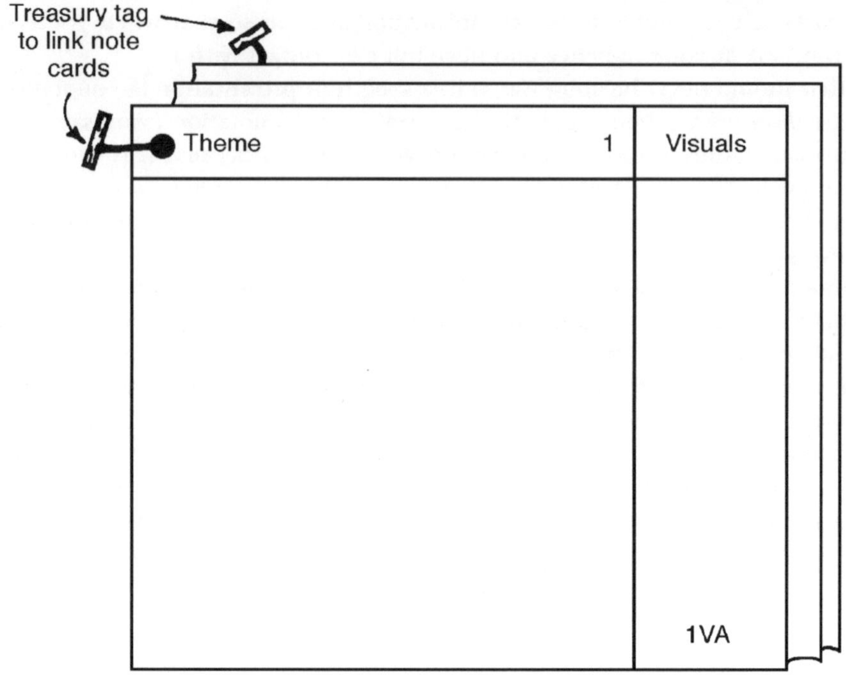

Figure 2.3 Assembling your notes

Preparing yourself

Having prepared your notes, you must now prepare yourself. Remember that only those without talent do not experience stage fright before making a presentation. A degree of anxiety before getting to your feet to speak is natural: it shows a concern for your audience and a determination not to disappoint them. You can reduce your anxieties by self-preparation. Frank de Angeli, sometime Vice Chairman of Johnson & Johnson International and an old colleague of mine, once gave his view on rehearsals to his managerial staff in the UK:

> Any manager calling a meeting should pay those who are to attend it the courtesy of not only preparing it thoroughly, but also preparing it beforehand before a group of his management colleagues.

I once actually witnessed him asking a manager to show him his diary of future commitments. In it there was a staff meeting. 'But', said de Angeli, 'I cannot see any date set aside for rehearsing this meeting. Why not?' 'Well I haven't time for that.' 'Then', replied de Angeli, 'cancel that meeting. I will not have our staff insulted by a manager who is unprepared.' Good presentations are neither rote-memorized recitations nor word-for-word readings.

(1) Read your presentation aloud to yourself to familiarize yourself with the content. Pinpoint any words that need careful pronunciation such as a person's name or a place.

Don't hesitate to replace a word that does not come off the tongue easily by another that does.

(2) Use a tape recorder to record and play back your presentation. From the point of view of timing it is useful to remember that we speak at about 130–150 words a minute. Another useful guide is to have on each of your cards an outline that represents at least five minutes' material.

(3) If possible, rehearse at the place where you are to speak so that you experience what it will be like.

(4) Ask some friends to your rehearsal to give you their constructive criticisms. If you can, also arrange access to a recording camera and closed-circuit television, then your rehearsal can be even more thorough. Not only can you obtain other people's assessments, but you can assess yourself. To aid your assessment, a presentation evaluation chart is illustrated in Figure 2.4.

(5) Beware, during such a rehearsal, of last-minute panic and a desperate search for something humorous or a funny story to leaven what may sound like a boring presentation. Inserted humour is always risky. My advice is to avoid trying to be humorous. Much better to think of all the ridiculous, bizarre, or embarrassing situations you have experienced and draw upon this material to tell a story or joke against yourself. You will be agreeably surprised at how effective this can be.

Speaking to foreign and mixed audiences

When you find yourself speaking to an audience many or all of whose first language is not English, slow your delivery down to about half the speed you would use when talking to an English audience – from 130–150 words a minute to about 60–90 words a minute. Remember that although non-English members of your audience may speak English, many of them, all the time you are speaking, will be mentally translating what you say back into their own language. This difficulty becomes even greater when you need to convey complex information. Slow down. Don't worry if what you say sounds ponderous or pedantic. Better this than to gabble away so that little of your presentation is understood.

An outstanding example of the pace and delivery needed when talking to mixed audiences across the world was provided by Ian McDonald, the Ministry of Defence official responsible for issuing statements to the world's press and television reporters during the Falklands crisis of 1982. He achieved national fame for the way he handled those press conferences. In slow and measured words, chosen with meticulous care, his communication objective was to present the facts as they became known, in language that would be clearly understood and about which there could be no possibility of hidden meanings, or alternative interpretation.

Another factor to watch when addressing mixed audiences is the danger of using slipshod English, slang or words with no exact translation or which, when translated, have a different meaning. Here are some examples of all three:

Slipshod English	*When you mean to say:*
'Let's wrap it up'	'Let us finish'
'No way'	'Not possible'
'Curtains'	'The end (for him or her)'

PRESENTER:	
SUBJECT:	
ASSESSOR:	

1 Beginning

(a) First impression

Dress	Immaculate		Acceptable		Inappropriate		Untidy		Comments:				
Facial expression	Pleasant		Enthusiastic		Serious		Nervous		Miserable		Comments:		
Stance	Upright		Still		Leaning		Moving		Aggressive		Humble		Comments:
Voice	Audible		Inaudible		Firm		Hesitant		Lively		Monotonous		Comments:
Mannerisms	Acceptable		Irritating		Appropriate		Inappropriate		Comments:				
Overall impression	Confident		Organized		Unprepared		Ill at ease		Comments:				

(b) Introduction of subject

Attention obtained by	Statement		Question		Visual aid		Dramatic		Story		Other:	
Theme stated	Yes		No		In his/her terms		In your terms		Comments:			
Rapport	Obtained		Not obtained		Mutual interests		Compliments		Sympathy		By personality	
Attitude	Confident		Nervous		Helpful		Critical		Apologetic		Enthusiastic	
Interesting	Yes		No		Comments:							

Figure 2.4 Presentation evaluation chart

2 Middle	Yes	No	Comments
Theme developed			
Points understandable			
Points attractive			
Points proved			
Objections anticipated			
Agreement on points			
Attention held			
Has first impression altered?			

3 End	Yes	No	Comments
Theme re-stated			
Points summarized			
Commitment requested			
Commitment obtained			
Did final impression match first impression?			

4 Summary

(a) What were the effects on you of this presentation?
(b) What caused these effects? Be specific
(c) What action do you recommend to reduce the unfavourable effects?
(d) What action do you recommend to maintain/improve the favourable effects?

Figure 2.4 (continued)

Slang	
'Put the screws on'	'Put pressure on (someone)'
'In the nick'	'In prison'
'Everything is A.O.K'	'Everything is alright'

English words	*Their translation into other languages*
Manager	In French it means a director, in Italian and German it translates differently again

Eventually (used to mean finally)	When used in a French sentence, and referring to someone doing something, it would translate as: 'He may have done it or he may not have done it.'
Blanket (proposal)	Whereas we use such a word in the context of 'a general approach' or one covering all possibilities, there is no exact translation of this word in most European languages.

It is impossible to exaggerate the value of thorough preparation. It makes the difference between mediocrity – or even catastrophe – and triumph. Figure 2.5 provides an outline for effective speaking that incorporates the guidelines offered in this chapter.

1 The title of my talk is ....

2 The objective(s) of my talk is/are * ....
* ....
* ....

3 Who am I talking to? ....

4 What are their NEEDS? ....

5 OPENING WORDS ....
- how am I going to gain their ATTENTION? ....

6 MIDDLE .... Do I need charts, props, visuals?
- how am I going to maintain their interest? ....

7 ANTICIPATE OBJECTIONS ....
- what objections will there be? ....
- how will I identify, handle/answer? ....

....
....

8 CLOSE How will I end my talk?
- by summarizing? ....
- with a story? ....
- by 3-step formula? ....
- with a poem? ....
- by asking for action? ....
- by assigning a task? ....
- with alternatives ....

REMEMBER TO:
- smile
- maintain eye contact with your audience
- start well, get better as you end
- end on a high note
- write your talk as a check list
- keep your visuals few and simple
- stand still

Figure 2.5 Effective speaking plan

Preparing a speech in note form

In the year 2000 an international pharmaceutical company invited me to open their annual gathering of senior managers in Nice. My brief was to set the scene for the programme and to stimulate the minds of the participants from the outset.

The following pages reproduce the original draft of the speech I gave, the marginal notes I made on it and the cards I prepared from this draft and which I used as notes when I actually spoke.

This speech contains all the essential ingredients for an effective presentation. As you read through it, first notice that, at the beginning I state a *theme* which is linked to the needs of those attending the programme; from this theme, I develop *three questions*; in the middle I *answer* these *three questions* and *support* the answers with *visual aids* to increase the impact.

Then, at the end, I *summarize* by reminding my audience of those three questions and I present an example which I hope provides a vivid picture about leadership and a piece of string. I close by *re-stating my theme* so as to leave in everyone's mind no doubt about what I believe is the cornerstone of successful management.

LADIES AND GENTLEMEN	**Comments**
Many of you will be familiar with the saying, attributed to Bernard Shaw: 'He who can, does; he who cannot, teaches.' And probably, like me, when you first heard it, you said to yourself, 'How true. I wish I had said that.'	**Say slowly** **Audience rapport** **Pause**
But when I heard it a second time, I said rubbish and, for two good reasons; *first,* as your chairman mentioned when he he introduced me, I started my adult life teaching English at an English independent boarding school, so as you can imagine, I was biased. But *second,* and more importantly, it seemed to me then and still does now, that if you can't train and teach, you can't manage anyone.	**Emphasize dramatically**
That, then, is the central theme of what I want to say to you all at the start of this programme which aims to provide you with some firm foundations upon which to consider your management job and how you are tackling it.	**Theme** **Your needs**
In a splendid biography of Bernard Montgomery, the famous British Army leader in the Second World War, the author Nigel Hamilton wrote that the core of Monty's success as a commander was this simple creed:	
'He picked those with leadership qualities and trained them to become effective. That was the underlying principle of all training, the instruction of the leaders *before* they trained their men.'	**Read out slowly** **Stress**
Incidentally, quite a few high-ranking officers in the British War Office thought that Montgomery was mad for preaching such a philosophy!	**Pause afterwards**

IF YOU CAN'T TRAIN YOU CAN'T MANAGE ANYONE	**Visual aid 1**
Training is the magic ingredient that binds leadership and management	
That is my theme. I propose to use it to discuss with you three questions:	**Theme**
(1) What do we mean by the term management? (2) What must a manager know and be able to do? (3) What are the key tasks of management?	**Visual aid 2**

Let us look at the first and most fundamental of these questions:

WHAT IS MANAGEMENT?
Over the last forty years, I have spoken to more than 50,000 directors and managers at all levels of responsibility in companies as large and global in their reach as American Express, IBM, GlaxoSmithKline, Shell and Unilever and in all parts of the world. It is still the exception rather than the rule for any man or woman to be told before or after being appointed to a management job, *what* in simple terms the job of management is in many of the organizations I have mentioned. My own entry into management fifty years ago was shrouded in the same mystery. Successful management can be stated simply as:

The ability to achieve planned objectives through the efforts of other people – NOT FOR THEM.	**Visual Aid 3** **Repeat last three words**

Note well those last three words NOT FOR THEM. When you are good at doing something yourself, it is easy to slip into the philosophy as a manager: 'If a you want a job done well, do it yourself.'

Let us consider some of the assumptions that are contained in this simple – yes deceptively simple – but I hope acceptable description of your job, for example:	**comments**

- it assumes, doesn't it, that 'your people', your personnel, want to help you to achieve your planned objectives?
- it assumes that by doing so, they will at the same time meet and achieve their own personal needs and aspirations.
- it assumes that you do of course know what those very individual, very personal needs and aspirations are.

Pretty BIG assumptions, aren't they?	**Rhetorical Q then pause**

WHAT MUST A MANAGER KNOW AND BE ABLE TO DO?

He or she must be able to do four main things successfully:

PLAN ORGANIZE MOTIVATE CONTROL	The efforts of others so that he/she can achieve	planned business objectives	**Visual aid 4**

Each of these four headings provides the basis for individual sessions I understand you are going to study in depth later in this programme.

Incidentally, for the word 'business' above substitute 'military', 'civil' or 'pastoral' and the original three questions about successful management apply just as well to military, civil service or church management.

WHAT ARE THE KEY TASKS OF MANAGEMENT? **Rhetorical Q**
Once you have planned your business objectives, and recruited the staff who, through you, are going to achieve them, you still have in the main uncut human diamonds.
They will not sparkle in their jobs unless you polish their skills. **Comments**
And how, you ask me, am I as a manager to do that? *By training, coaching and developing them* – the heart of my talk and my theme. **Rhetorical Q** **Stress**

So the key tasks to which you as managers should devote most of your time are:
(1) Training *new* staff on the job that they are to perform. **Visual aid 5**
(2) Training and developing *experienced* staff on the job they perform and *on a regular basis*.
(3) *Appraising* individual performances.
(4) *Motivating* your staff to achieve their job objectives.

Later in this programme you are going to be shown, and then develop your skills in carrying out, these key tasks of management. Remember: no one can really enjoy ajob unless he or she can do it competently. And who at the start of their career holds the key to that door? YOU DO. **Rhetorical Q**

This is where you earn your salary as a manager. When I was first made a manager at the tender age of 27, I became responsible for twelve salesmen whose average age was 57 and none of them was in the least bit interested in what I had to say except that more than one in that group thought that he should have had my job!
It took me over a year to win over those two and I had to do it by finding out how I could help them where they needed it, not by throwing management theories at them.

Questions invited here

Final attention getter before conclusion

Before I summarize, you probably have initial questions you would like to ask me about some of the contentious issues I have discussed, indeed some of the statements I have put forward.

Concluding remarks

Summarize on points

First in this short talk today, I have given you my overview of what the term MANAGEMENT means: it is about getting results through people. What those people become and what they are capable of doing depends on what you do with and for them **1**

Second I have discussed WHAT MANAGERS MUST KNOW AND BE ABLE TO DO **2** and in particular, WHAT ARE THE KEY TASKS OF MANAGEMENT? **3**

At the outset of my talk, I mentioned that there is a magic ingredient that binds management and leadership. Leadership, you know, can be compared to a piece of string. If you pull a piece of string, it will follow you wherever you wish to go. But if you try to push it, it will either go nowhere at all or all over the place.

Slowly and with emphasis

Now people are the same when it comes to leading them. Leadership depends upon your ability and skill in making your staff want to follow you. But, I can hear you saying to yourselves, how will I do that? You will get the best out of people by helping them to do their best by showing them HOW to do their job; HOW to develop their skills; HOW to get results through your knowledge and skills as a TRAINER and DEVELOPER.

Rhetorical Q then pause by stating the 3 HOWS

But, to return to the theme with which I started and will now end,

IF YOU CAN'T TRAIN YOU CAN'T MANAGE ANYONE.

Theme. End on a loud note on final sentence

The cards in Figure 2.6 (on pages 35 and 36) show how this talk can be summarized in note form and it was from these cards that I delivered the talk.

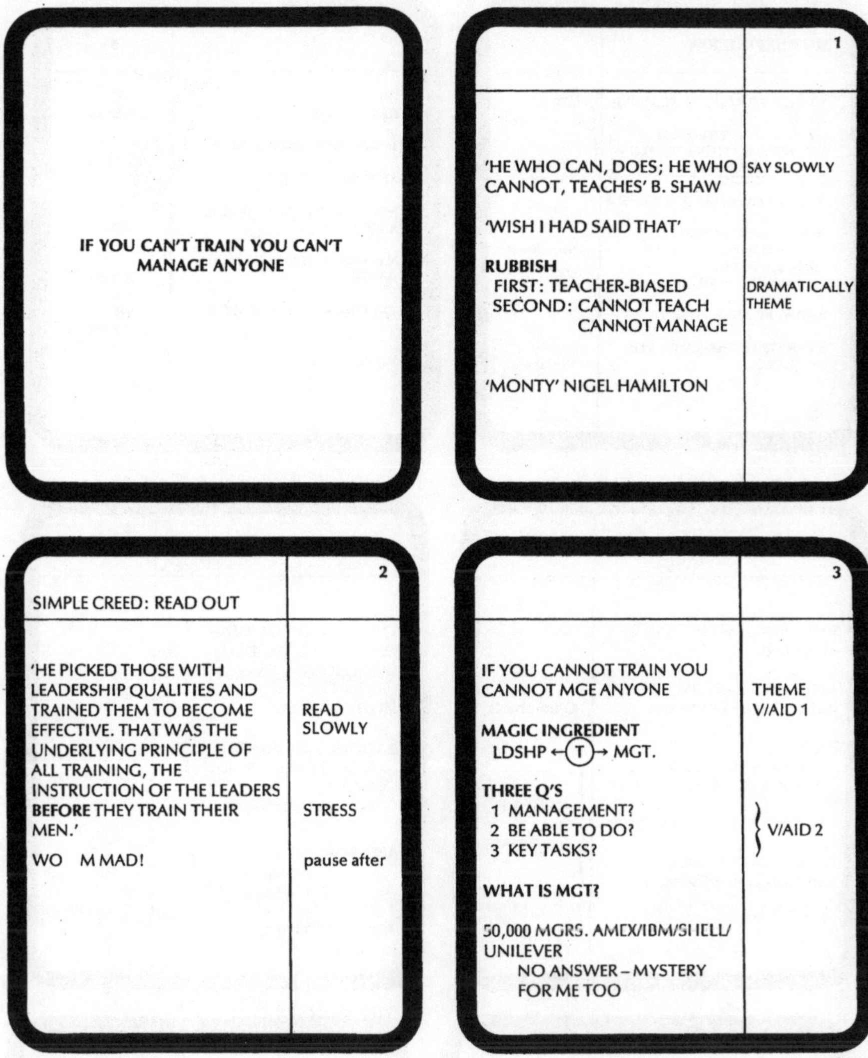

Figure 2.6 Talk summary cards

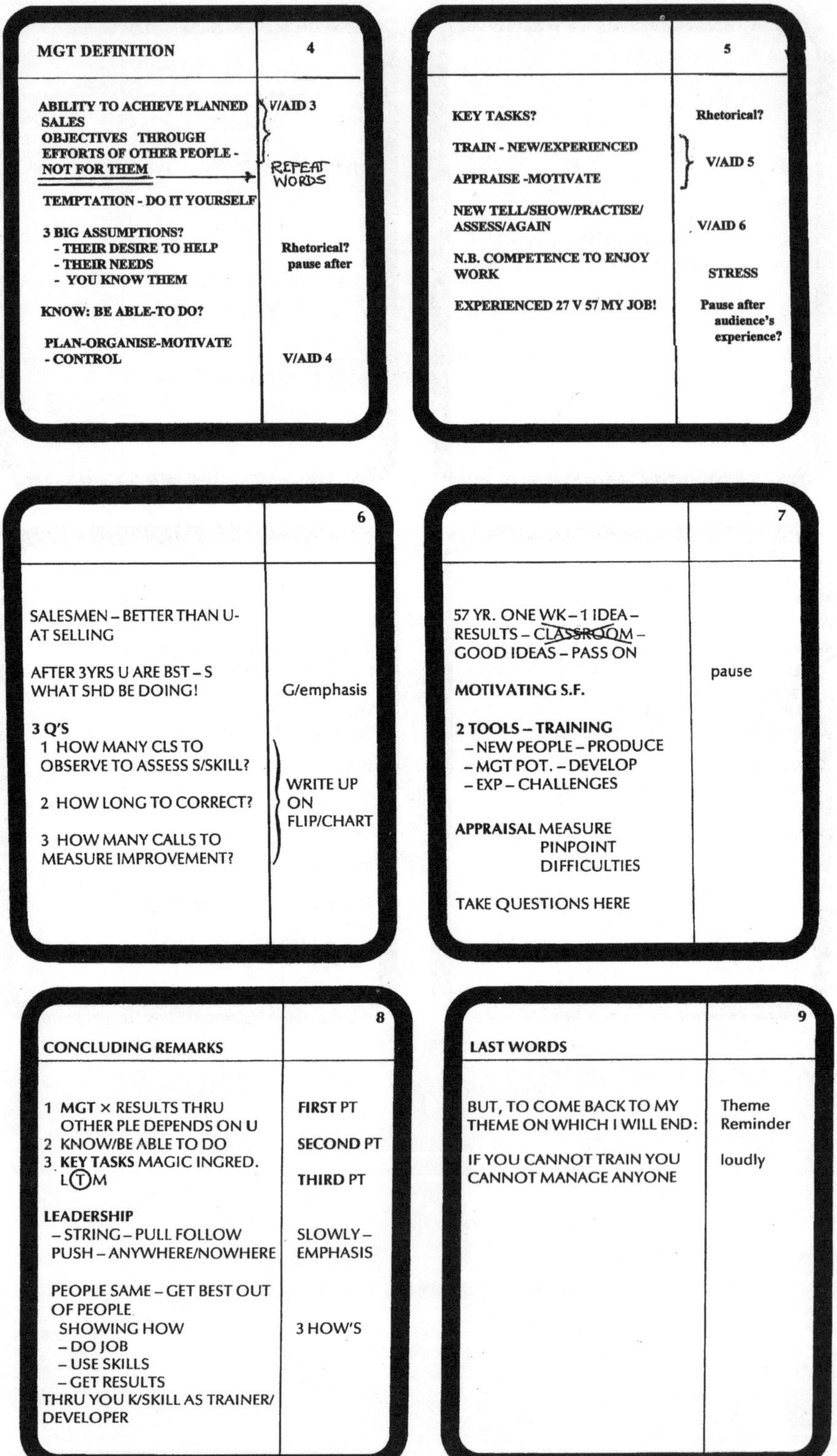

Figure 2.6 (continued)

Let me end this chapter with a set of simple dos and don'ts for speakers to bear in mind.

Dos and don'ts for speakers

DOS

have your clothes pressed
dress as the group does
look like an expert
stand with your back to a wall or curtain
be yourself
write your speech in checklist form
smile from time to time
talk more loudly than normal
maintain eye contact
face your audience
stand still
stand erect
lean forward slightly on your toes
leave your spectacles on or off
use a variety of gestures
tell 'em what you're going to tell 'em
then tell 'em
end by telling 'em what you told 'em
end on a high note
keep visual aids covered until you need them
remove visual aids when they have served their purpose
finish before you are expected to

DON'TS

write your speech as an essay
read your speech
talk to your notes
have distractions behind you
talk to your visual aids
talk to the blackboard or screen
walk up and down
lean on the lectern
fidget
fiddle with your clothes
smoke
use the same gesture repeatedly
compete with distractions
compete with your own material – if you hand something out to be looked at stop talking until everyone has examined it
wear clothes that distract from what you are saying
fidget with your notes
overrun your allotted time

CHAPTER 3 *Beginning your Presentation*

In the previous chapter we recommended a three-part structure as the best basis for planning your presentation. In this chapter and the two that follow we examine each of those parts in detail. The first thing to observe is that there are specific objectives belonging to each part. Similarly, each part corresponds to some of the seven points that make up the listener's thinking sequence described in Chapter 1. The beginning of the presentation, for example, corresponds to the first two steps in that sequence:

(1) *I am important and want to be respected.*
(2) *Consider my needs.*

Your three objectives and how to achieve them

In this opening phase of your presentation your objective is threefold:

(1) to gain the undivided *attention* of your audience
(2) to build *rapport* with your audience
(3) to state the *theme* in terms of the *needs* of your audience.

Let us now examine each objective in turn and consider how to achieve them.

OBJECTIVE 1 TO GAIN THE UNDIVIDED ATTENTION OF YOUR AUDIENCE

Even before you start speaking, you will make an impact on your audience through your *appearance* and *manner*. However unfair it may be, audiences tend to make quick judgements based on first impressions. So make sure that you:

(a) *Stand up straight* in a comfortable stance with your feet slightly apart so that your body is well balanced. Push your chair well back behind you to provide yourself with ample space to move about freely, without constraint or worry.
(b) *Dress comfortably, neatly and appropriately* for the occasion, avoiding anything that might distract the audience from what you are about to say. Remember the occasions when you have failed to give your full attention to a man dressed in outlandish attire or a woman with glittering brooch and bracelets.
(c) *Take a good, deep breath* so that you do not cough up or splutter out your opening words. On film sets, for example, it is quite usual for the director to ask members of the cast about to perform to have a good throat-clearing session first.

(d) *Pause,* and don't start until there is complete silence. This is in itself a most effective attention-getter. It means that the whole audience is concentrating on you and what you are about to say.
(e) *Look around your audience* to establish eye contact with everyone present.
(f) *Learn your opening sentences by heart* so that you can catch and hold your audience's attention by eye contact from the first word and don't lose it by burying your head in your notes.
(g) *Talk more loudly than for normal conversation.* You must make instant impact when you speak, so pitch your voice much more boldly than you would in conversation. Make it loud enough to be heard clearly by the person seated furthest away from you. If he or she can hear you, then so will everyone else. There is no point in asking your audience 'Can you hear me?', as speakers often do, because, for some strange reason, few people will admit that they cannot hear you even if is true!
(h) *Avoid distracting mannerisms.* Empty your pockets of all loose coins so that if you put your hands in them there will be nothing to jingle, an infuriating habit of some men. Likewise, women should avoid such jewellery as wrist bangles that move about noisily when you gesture with your arms. Keep still. Do not walk up and down like a caged animal. Beware of fixing your gaze on a particular person in the audience for too long or looking at only one section of the audience.

The first few minutes of any presentation are the most difficult. They are also the most important, because not only what you say but how you say it can either win or lose your audience there and then. *Drama, curiosity, a story, a checklist* or *a series of questions* are all useful devices for capturing audience attention at the beginning.

Dramatic openings

Accidents and crime prevention are not obvious subjects for drama, but they can be used effectively. Here is how one speaker began his a talk on this subject: 'There are just over a hundred people in our audience this evening. In the next twenty minutes, while I am speaking to you, over one hundred robberies will take place in this city tonight.'

Another excellent use of drama was to demonstrate the value of training: 'Yesterday a plane in which my husband was one of the 280 passengers crash-landed at Heathrow airport. My husband and his travelling companions all owe their lives to the thousands of pounds spent on training the pilot of that aircraft who knew, in that moment of crisis when the undercarriage failed to operate, how to bring the aircraft in to land with the greatest chance of saving the lives of those on board. In that moment, his training paid out a massive dividend.'

A series of questions

Questions can provide both a thought-provoking start to a presentation and a framework for it. For example: a medical director giving a talk to senior members of her specialist management team:

> There are three questions I would like to use as the agenda for our discussion about osteoporosis: First, what can our company do to encourage the women most vulnerable to osteoporosis, the over 55-year-olds, to have a bone density scan? Secondly, how can we secure the co-operation of the 40+ other pharmaceutical companies with products for the treatment of osteoporosis to join in a campaign? And

thirdly, how do we make sure that the NHS will support the GPs who will have to deal with the referrals resulting from probably three million scans?

OBJECTIVE 2 TO BUILD RAPPORT WITH YOUR AUDIENCE

A part of the secret of any successful presentation lies in the feeling of oneness between you and your audience. Your audience must warm to you and feel that you are sincere and believe in what you are saying rather than just putting on an act – a criticism levelled at politicians today, hence their unpopularity. Depending on circumstances, one or more of the following will help to build rapport:

(a) Compliments

If your audience belongs to a company or institution that has achieved something notable, you can start your talk by expressing admiration for it. But such compliments must be sincere. Here is an example of a guest of honour speaking at the twenty-fifth anniversary dinner of a large clinical research agency: 'For all of you here tonight, this is a special year. You have been in business for twenty-five years. Some cynics might say 'so what?' I would answer them, that it is no mean achievement to have reached such a milestone in an industry where keeping talented staff alone is itself a challenge, and to have motivated them to achieve the goals you have also delivered for clients like my company is outstanding. Congratulations'.

(b) Mention a common interest

If you and your audience share an interest, it can be mentioned. For example:

> Ladies and gentlemen, we all share an interest in pharmaceutical research and development, despite working for competing companies. So I thought that it would be appropriate to begin this debate by quoting from a talk given by the late Robert Johnson, the founder of Johnson & Johnson, in the early 1900s, which is as pertinent today as it was then: 'The [R&D] department is not conducted in any narrow commercial spirit, and not kept going for the purpose of paying dividends, or solely for the benefit of Johnson & Johnson, but with a view to aiding the progress of the art of healing the sick'. I cannot think of a better reminder for us all of what our work is all about.

(c) Demonstrate your own competence – without boasting

> Let me, as your programme director, tell you a little about myself: managing director of a leading healthcare management consultancy, I have a degree in pharmacy and started my commercial career, like all of you, in the pharmaceutical industry, first as a medical rep, then running clinical trials. Finally, as a marketing manager, I had to launch a complex and expensive prescription product into the cardiovascular market worldwide. So I share with you the experiences of having sold, product managed and marketed ethical prescription medicines.

Always remember: *radiate enthusiasm* as you begin – it will make your audience enthusiastic. Your tone of voice, an occasional smile, the expansive gesture, all can bring a warmth which is infectious to your presentation.

OBJECTIVE 3 TO STATE THE THEME IN TERMS OF THE NEEDS OF YOUR AUDIENCE

Relating your theme to the audience's needs is important because it sets the tone of the whole presentation and provides them with a point of reference. For maximum impact the theme should be stated at the start, wherever possible, in terms of audience needs. If, for example, your aim is to build a bridge between medical and marketing departments, you could begin: 'Our industry is becoming fiercely competitive and so we have a vested interest in the success of our marketing colleagues. It is about how we and they can work together that I want to talk about today'. Contrast such a beginning with the far less appealing: 'I'd like to talk about my ideas on marketing today'. In short, unless your audience understands what the subject *means* to them, they will lose interest very quickly.

If your presentation is going to cover several points, it is helpful to mention them at the beginning so that the audience knows what to expect from you and has a structure they can follow. For example: 'First, I will tell you about the main reasons for this study. Secondly, what the main findings are. Thirdly, the conclusions we arrived at. Finally, the main recommendations upon which you can base an action plan for your individual Primary Care Trusts'.

Sometimes you may know very little about the needs of your audience. In that case you might begin your presentation with a series of open-ended questions until you obtain agreement on what they are looking for. These questions, written up on a flip-chart or overhead projector transparency, can provide the structure for your presentation. For example: 'You have asked me to attend this meeting to describe how we could co-operate with your company on a joint research programme to economize on costs. First, could you tell us why you think we are a best match for you? Secondly, how open are you prepared to be with us about your findings up to now? Thirdly, what are you good at doing, and what not so good?'

The aim of describing these various types of opening has been to show you, through examples, how to begin a presentation so that you can achieve the three objectives, which you will remember are:

(1) to gain the undivided *attention* of your audience;
(2) to build *rapport* with your audience;
(3) to state the *theme* in terms of the *needs* of your audience.

Conclusion

The opening stage of a presentation sets the scene for everything that follows. You, the speaker, want the audience to have confidence in you, and you want the undivided attention of each one of them. You want to establish with your audience some objectives. You want them to believe that you have something to say that they will want to know and hear about. To achieve these things, you must appear confident, enthusiastic and keen to help them. Above all else, remember the golden rule when preparing your opening words: *If you don't strike oil in the first three minutes, stop boring.*

CHAPTER

4 *Conducting your Presentation*

Having begun effectively by capturing your audience's *attention*, building *rapport* and identifying a *theme* related to the needs of those listening to you, you will now move on to the middle of the presentation. Like the opening stage, the middle corresponds to some of the seven steps in the listener's thinking sequence, in this case:

(3) *Will your ideas help me?*
(4) *What are the facts?*
(5) *What are the snags?*

Your four objectives and how to achieve them

Your aim in this middle phase will be to maintain the standard of your opening. To do this you will need to achieve the following four objectives:

(1) to present your ideas in detail
(2) to have each point you present accepted
(3) to keep your audience's attention
(4) to prevent objections or handle them satisfactorily.

Let us now look at each of these objectives.

OBJECTIVE 1 TO PRESENT YOUR IDEAS IN DETAIL

By this time your audience will know the theme of your presentation and, if it has been stated in terms of their needs, they will now expect to be told how you propose to meet them. For the sake of clarity and to aid audience acceptance, it is best to take *one point at a time* and deal with it before moving on to the next. This can be done in two ways, depending on the subject matter.

Either:

(a) take each of the audience's needs in turn and present your ideas for meeting that need. For example, a speaker appealing for support to buy a specialist piece of medical equipment for a cottage hospital in an expanding town: 'Let me deal first with your concern about the provision for bone-density scanning at your local cottage hospital to cater for the high number of women aged over 55, a need that your research has shown is very important. Now there are at least three possibilities we can explore: first . . .'.

or

(b) take each of your ideas in turn and show how it meets their needs. Developing the example we have just used in (a): 'One idea I would like to suggest is co-operation between three local hospitals, since a larger catchment area would better justify the use of such an expensive piece of equipment. The volunteers from three areas would also be able to raise the necessary funds faster.'

When the subject permits, structure your presentation around the needs of your audience, because it is those that are uppermost in their minds.

OBJECTIVE 2 TO HAVE EACH POINT YOU PRESENT ACCEPTED

Acceptance of your points depends upon their being understood and seen to be of value in that they will produce a desirable result known to be valid and agreed.

(a) *Understanding* can be achieved by:

(i) Using language familiar to the audience and avoiding jargon, technical terms and unfamiliar words. For example:

'about' rather than *'in the region of'*
'soon' rather than *'in the not too distant future'*
'reduce' rather than *'effect a reduction'*
'question' rather than *'query the status of'*
'happen' rather than *'eventuate'*
'split' or *'division'* rather than *'dichotomy'*.

As George Orwell, the author of *Animal Farm,* once said: 'Never use a foreign phrase, a scientific word, or a jargon word, if you can think of an everyday English equivalent.'

(ii) Explaining your ideas by using similes, or going into detail, for example: 'This scheme which you are considering is almost identical to one that was adopted last year by a neighbouring town. Let me indicate the strong resemblance between the two.'

(iii) Using actions or gestures. Always make sure that an action or gesture does help to communicate the point you are making. To extend both arms to indicate massive size will be effective if you have not used this gesture before. It will fail in its objective if you have been waving them about the whole time you have been speaking. Leaning forward on the desk or lectern from which you are speaking and then stating a serious point can be most effective if your audience has so far not seen you do this.

(iv) Giving demonstrations. Nothing directs the attention of an audience from one point to another so surely as a physical demonstration, or showing a piece of equipment. But always be prepared. Practise in advance, not once but several times, what you propose to show. Take your time; tell your audience what you are going to do before you start, and give your reasons for doing it. I have a vivid recollection of an accountant demonstrating to a group what gives money value. He strode into the room where he was to speak carrying a large sack over his shoulder. Then he asked two members of the audience to check that all the doors in the room were locked from the inside.

Assured that all the doors were locked, he began:

> In this sack is £100,000 in used £10 notes. I am going to empty them all onto the carpet like this (he then did just that). Now any one of you can help yourselves to fistfuls of these notes. But as long as those doors leading from this room to the outside world where you could go and spend that money remain locked, these notes are nothing more than bits of paper without value. I want to tell you today what gives them value.

(v) Using visual aids. Like demonstrations, visual aids skilfully used can convey information more dramatically than the spoken word. Because they are so important, details of various types of visual aid and their advantages and disadvantages, together with some rules for their effective use, are set out in Chapter 6.

(b) *Acceptance* of the value of your ideas, plans or proposals is vital. To achieve acceptance: tell your audience what your ideas will do for other people in whom your audience are interested, say, their relatives, friends, superiors, professional colleagues, customers or shareholders. For example:

> As consultants in this hospital you will be interested in the easy access to patient histories that this computer provides. And of course, so too will be the GPs who send their patients to you for a second opinion.

Your audience wants to know what your ideas will do in terms of *what they want done*. If their prime concern is to be certain that your products or equipment can be operated easily by semi-skilled labour, it is pointless to emphasize that it fits into a small space, thereby increasing production per square foot! Their unspoken reaction will be either 'So what?' or 'That's not my main problem'. If you want your ideas or proposals to seem desirable to your audience, you must select the results they will get that fit in with their needs, and arrange the sequence with one result logically leading to another so that, eventually, their needs will be met. Here is an example concerning administrative savings:

> This idea would result in a reduction by two hundred tons in the amount of paper this group of hospitals would need to print medical records. Secondly, there would be a saving in administration costs of £175million a year – and what would that mean in release of funds for direct patient care? A massive amount.

(c) The *validity* of your points may be questioned, albeit mentally, by your audience, especially if your ideas or proposals are new or seem revolutionary to your listeners. Such points you can prove to be valid by quoting examples of other cases where they have worked. But when you give these examples, or when you refer to third parties who have adopted your ideas and benefited from doing so:

(i) don't start with such references; instead use them to support arguments you have already made;
(ii) ensure that the individual, group, company or people to whom you refer are respected by your audience;

(iii) ensure that the circumstances in both cases are sufficiently similar to make your point acceptable, rather than to risk the response 'Their situation is quite different from what we have at this practice';

(iv) tell your audience the desirable end-results the third party you quote obtained. For example, the managing director of a company that manufactures visual aid equipment is presenting the case for a hotel group to invest a considerable amount of money in new stands, screens, speakers, projectors and so on.

> Now you may all be saying to yourselves 'That sounds fine but he has a vested interest in persuading us. How do we know that this is appropriate for us at this residential training centre for our medical sales force as well as our management development?' Let me refer to two hotel groups, MacDonalds and the Marriott Hotels. First, like you, they cater for a very large percentage of business conferences and training meetings, some staged by your company. Secondly, in the last two years, they have attracted new business from companies that in the past have preferred their own in-house training facilities. So you can see that, whether you install this equipment here, or balance it with continued occasional use of hotels, your training people will be using equipment with which everyone is familiar and comfortable.

(d) *Agreement* on the part of your audience is not always visible. Blank silence can imply agreement, disagreement, bewilderment, boredom or even an audience capable of sleeping with its eyes open! Here in the middle of a talk or presentation, you need to seek evidence of agreement – or otherwise – on each point before moving on to the next one. How do you do this? You can check reactions by:

(i) constant observation of their facial expressions to see whether they are listening to you, nodding at what you say or vigorously but negatively shaking their heads. If you think about it, nodding your head and at the same time violently disagreeing with what a speaker is saying is quite a difficult thing to do. At a Marketing Society conference one of the key figures on breakfast television was invited to give a forty-minute presentation on the commercial benefits that his programme would offer his audience of more than 700 big advertising spenders. In less than ten minutes, he had so bored them that nearly everyone had picked up one of the free magazines on every delegate's chair and was reading it. Did he stop speaking and throw out a series of questions to assess attitudes? No. He just carried on with his head buried in his notes, determined to finish what he had started!

(ii) asking questions linked to a request for a show of hands if their facial expressions create doubt. For example, 'Could we have a show of hands. How many of you in this audience think that investing in advertising at this hour is commercially worthwhile? Now how many don't?'

OBJECTIVE 3 TO KEEP YOUR AUDIENCE'S ATTENTION

It is in the middle of a presentation that audience attention usually declines. This dip was illustrated in Figure 2.2 on page 23. The longer the presentation, the greater the danger of losing attention in the middle. To keep it high you need a series of carefully timed 'attention-getters' as shown in Figure 4.1. In detail:

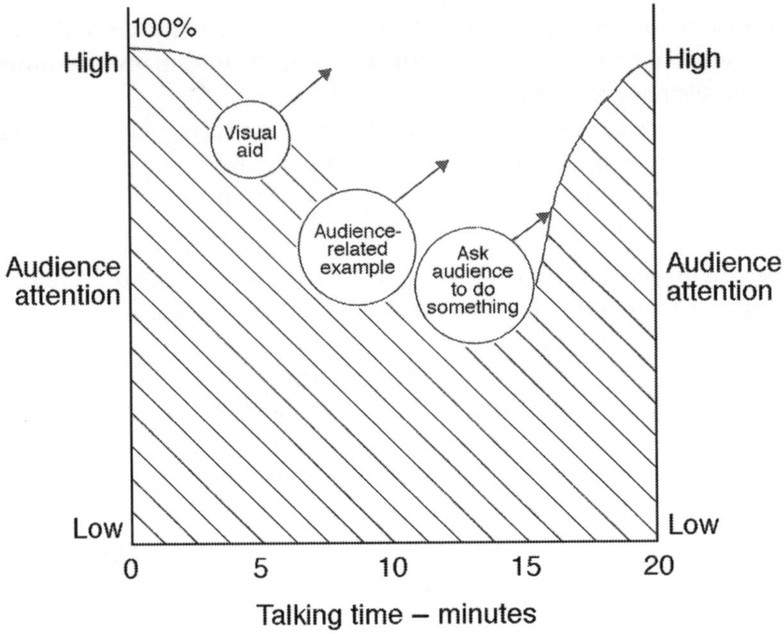

Figure 4.1 Attention-getters

(a) Keep telling your audience what your ideas mean to them, using on occasion the rhetorical question which does not need an answer but is a most effective means of emphasis. For example, the marketing director, his product manager and two medical representatives of a company with a product in the influenza vaccination market, are rehearsing before the company's medical director and her staff a presentation to made to a PCT* to supply sponsored nurses during the busiest period, from October to the end of November, for vaccinating patients against influenza in three group practices where there are large number of elderly patients eligible for vaccination.

> You are perhaps asking yourselves the question as a PCT, if we accept this offer of sponsored nurses, will it benefit the three practices we have identified? We can assure you that you and the practice staff will all benefit, for the following reasons: First, the sponsored nurses will be able to shoulder the responsibility for carrying out the whole 'flu vaccination programme and recording the patients vaccinated in the computer records. Secondly, you will be able to bring forward your annual vaccination programme in these practices which in the past has begun at the end of October to the end of September. And thirdly, the provision of these sponsored nurses will release the full-time nurses in these group practices to deal with the increased calls on their specialist skills at this time of year.

(b) Keep their eyes occupied by using visual aids, demonstrations, and so on.

* PCT equivalents: in Scotland, same; in Wales, Local Health Board (LHB); in Northern Ireland, Local Health & Social Care Groups (LH & SCG).

(c) Where possible give the audience something to do, for example, 'Let us find out what our immediate neighbours think about this issue. Please ask the person sitting next to you what his or her answers are to these questions: First; second; third.

(d) Quote stories or examples and, if you feel confident about doing so, invite members of the audience to share some of their experiences by standing up and giving them. If you do this, you must be prepared with methods of stopping what could become an endless talking match by saying, for example, 'Thank you for those ideas. I can take just one more, and then I would like to discuss another very curious development with you.'

(e) Maintain your own enthusiasm. This advice is easy to give but how do you generate and maintain enthusiasm? First, you must believe deeply in what you are talking about. If you don't, then why on earth are you on your feet talking at all? Your beliefs, your convictions must come from your heart. As Dale Carnegie put it: 'Talk about something you have earned the right to talk about.' Speak from your own experiences, tell stories that relate to you. They will be far more convincing than anyone else's. Avoid, if you can, using second-hand material.

A former member of my staff used to sit in on my talks noting down the various stories I told that were drawn from my own background. I witnessed him doing this at first hand on one occasion. After he had mis-told one of my stories, a member of the audience said, 'That is very interesting. What happened next?' And of course the speaker was nonplussed because it was not his story so he could not answer the question and lost face! And deserved to do so. Look at your audience from time to time and smile as though you are genuinely enjoying what you are doing. Once again, I say, if you are not, why are you speaking at all? Use dynamic, emphatic gestures. Remember you are your own most effective visual aid.

(f) Involve the audience where possible and appropriate by asking people to respond with a show of hands to a question or asking them to look at a particular visual aid and identify a particular feature. Or to take a pen or pencil and write down their reaction to something you have said. Then ask for feedback on what they have written down, using it to develop your talk further.

OBJECTIVE 4 TO PREVENT OBJECTIONS OR HANDLE THEM SATISFACTORILY

Before taking any action human beings instinctively consider the possible 'snags' or disadvantages to them that might arise as a consequence. This is true throughout our daily lives: we are constantly having to weigh up whether or not to buy this, whether we should go on holiday this year or spend the money on something the family needs even more than a break. In formal presentations, talks and speeches the audience is just as likely to think of objections to a speaker's ideas as an individual would do face to face with another person. The main difference is that objections are less frequently voiced in formal presentations. Perhaps because of this reticence, many speakers ignore the need to recognize not only that objections are bound to arise in the minds of the audience, but also that they must be dealt with satisfactorily and openly – and sometimes, perhaps, acknowledged as unanswerable.

Stand in the foyer of a public meeting place and listen as an audience files out at the end of a speech by a politician. You will hear such comments as 'He never answered that question, did he?' 'She was too scared to tell the truth about whether, if returned, this lot will raise prescription charges yet again, wasn't she?' Remarks like that suggest another set of communication failures.

Why do objections arise in the first place? Most objections are not inherent in the audience. People do not deliberately sit down and work out trick questions for a speaker. Objections are created by the speaker because:

(a) the needs of the audience have not been sufficiently explored in advance
(b) solutions to ideas are proposed too soon, and can appear slick or contrived
(c) benefits and technical details are not specific enough or related closely enough to the needs of a particular audience.

What can you do about objections? Two things; you can *anticipate* them, and you can *know how to deal with them* when they arise. It is important that, as far as possible, objections are considered in advance by the speaker and answers woven into his presentation. For example, suppose a speaker is presenting a case for spending money and knows that it is going to generate a widely shared concern about cost. The fact can be explicitly acknowledged: 'Some of you may feel that the investment of £50 per patient, multiplied by the number of elderly people likely to have broken bones, arms, shoulders and legs, is a high one for the NHS. But compare the cost of that preventive action with the hundreds of millions of pounds the NHS would save by not having to perform orthopaedic surgical operations. It's like investing in education: consider the alternative!'

If objections are expressed during a presentation as a result of the speaker's invitation to the audience to ask questions, the objector wants their views acknowledged by the speaker and answered sympathetically. It pays a speaker to handle such objections by:

(a) pausing after the objector has finished speaking. This gives the speaker time to think and also prevents the temptation to crush the objector with a clever or instant rebuttal;
(b) listening attentively to the question and even writing down the gist of it. Then, in a voice loud enough for all to hear, repeat the question so that everyone understands it. Members of an audience who raise an issue have not usually been trained in public speaking and their questions are often inaudible to others, particularly if the questioner is sitting in the front row;
(c) checking with the objector that you have understood them correctly;
(d) acknowledging that the objector has a point, for example by saying, 'Yes that is an important consideration, which I would have asked had I been in your position';
(e) answering the question by concentrating on what the objector wants to know.

If the objection is unclear to you, clarify by asking the objector to explain what they mean. Frequently this will provide you with another angle which can help you to develop a satisfactory response.

Another technique is to clarify the question with the objector and then refer it to the audience as follows: 'While I have some ideas on this subject, are there others who share the questioner's concern?' Never call the questioner an 'objector' which sounds confrontational. Then, 'What are your own views on this?' The responses from members of the audience can provide you with a valuable insight into what a range of people think. It can also endorse or qualify the answer you were about to give.

In many presentations, the middle is the part least remembered, often because speakers try to say too much and explain too many points. It is the part where attention is lost, where speaker credibility falls, where objections arise, and where audience rejection of the speaker and his ideas sets in. It is also where many speakers alienate the audience with an unguarded but damaging comment. If the beginning of your presentation has been good, then the middle is where it should become even better.

Remember to:

(a) take one thing at a time – ideally let your presentation be about one point;
(b) keep emphasizing what the point you make means to the audience;
(c) maintain their attention with visual aids, stories, questions, demonstrations, involvement and so on;
(d) conclude each point to the audience's satisfaction before moving on to the next one.

Dealing with questions

Some speakers are happy to accept questions from the audience as they arise during the presentation. But to be able to handle them spontaneously requires practice and much skill. Also, when a meeting is being chaired by someone other than the speaker, that person may not like the idea of questions being raised during the presentation, not least because of the problems it creates for him or her in controlling the meeting.

If you are prepared to take questions from members of the audience as part of your presentation, it is better to do it before you summarize and close so that you are in final control of the meeting and can thus ensure that you end on a high note. If you finish your presentation and then invite questions there is the danger of your presentation petering out inconclusively, unsatisfactorily and possibly so contentiously that everything you said is forgotten. You have probably been present when a questioner asks a question they know is unanswerable, the speaker stumbles in replying and the audience is left with the impression that the questioner has bested the speaker. That is not what you want, nor is it one of the objectives you set out to achieve when you agreed to speak.

Here are some guidelines on questions and how to handle them:

(1) Plan *when* and *how* questions from the audience will be handled and agree this in advance with the chairperson of your meeting, if there is one. If there isn't one, say to the audience: 'You will, I am certain, want time for questions, so I propose to take them just before I close.'

(a) *Questions taken during the presentation*
The advantage of handling questions in the course of your presentation is that this is when they have most relevance and meaning and provide feedback that shows how far the audience has understood what you have spoken about.

There are a number of disadvantages:

- they can disrupt your timing
- they may interest only a small part of the audience so that you risk boring or antagonizing the remainder
- premature questions can disrupt your sequence if you answer them immediately.

How to answer questions in this situation is a matter of personal skill in evaluating the question and the questioner quickly:

- is it relevant?
- is it covered later in the presentation? If it is say so, without being dismissive.
- is it necessary to the understanding of what follows?
- is it a good opportunity for reinforcing? 'Your question is timely because it serves to underline the point I have just been making. Thank you for raising it.'

(b) *After the presentation*
Avoid at all costs a question-and-answer session after your final summary. If you fit in a session beforehand, you can finish on a high note, minimizing the likelihood that negative attitudes will arise during questioning. You can, if appropriate, announce that you will be prepared to remain after the meeting to talk to members of the audience individually. This also deals satisfactorily and on a one-to-one basis with argumentative people who, if allowed on their feet during the presentation, can be disruptive and even cause serious damage to your effectiveness.

(2) Questions and questioners

(a) *The argumentative type*. Do not try to win the argument – it cannot be done. Generally this type of questioner is looking for recognition from the rest of the audience – so give it to him or her. 'That's a good point, can we go into it in detail after the meeting and thanks for bringing it up' or, 'That's a good question, thanks for raising it'. Answer quickly with a promise to the questioner to elaborate later, then resume your presentation.

(b) *The loaded question*. This is another recognition-seeker. It is usually unanswerable, like measuring the unmeasurable; in any case such questioners frequently have their counter-answer ready to turn the question back to the speaker.

(c) *The rambler*. Sometimes you can anticipate the question and immediately both ask and answer it. If it relates to what you have already said you can interrupt and refer back with emphasis to the relevant point. Sometimes you can defer the answer until after the presentation and discuss the question with the person who raised it.

(d) You *don't have an answer*. Admit it. You can win support by doing so. If you try to bluff you can lose credibility.

(e) You *need time to think*. Pause, then say one of these:
'Would you repeat or elaborate, please?'
'Good question, how do you feel about this?'
'Good, point made, let's think about it'
'Interesting question, how do the rest of you feel about it?'

Sometimes you can write the question up on a flip-chart before answering. That also has the effect of underlining the importance of the issue behind the question. The way you handle the question can be crucial to the atmosphere of the presentation and your acceptance by the audience. You need all your perception and patience to do it well.

The golden rule is: NEVER EMBARRASS A QUESTIONER.

CHAPTER 5 Ending your Presentation

No matter how long a presentation has lasted, no matter what the subject, *the audience expects it to end on a high note.* If it was good at the beginning and became better in the middle, then it deserves to be – must be, indeed – best at the end. It is better to start with a cap pistol and end with a cannon than the other way round. And if the presentation has not gone so well, the ending gives you one final chance to make a good impression.

The end is where all the threads need to be drawn together and all your presentational skills combined to produce a climax that leaves your audience *impressed, convinced* and *eager to act in your favour.* And of course it relates to the last two of the seven steps in the listener's thinking sequence:

(6) *What shall I do?*
(7) *I approve.*

Your objectives at the end

From the listener's standpoint your presentation has reached the stage where he or she is faced with a decision. Do I accept or reject what I have been listening to? In choosing between acceptance and rejection the members of the audience will concentrate on their own needs, personal or professional, and decide accordingly. If they are faced with several sets of ideas or proposals they will prefer those that they believe best meet their needs. If these needs have been satisfactorily met they will make a decision in the speaker's favour *I approve.*

So your objectives for the ending are:

(1) to re-state the theme of your presentation in terms of audience needs
(2) to summarize the main points of your presentation
(3) to close calling for a commitment based on the objectives of your presentation.

THE PSYCHOLOGICAL BARRIER

Many speakers feel uncomfortable at having to conclude a presentation or talk. They fight shy of asking for a commitment. This is very similar to the problem every medical sales representative faces at the end of a discussion about his company's products and what they can do for the GP's patients – how do I ask him or her to agree to prescribe them? An audience wants a commitment. Speakers fight shy because they are afraid of a rejection and prefer to leave issues undecided. Such an attitude is understandable, but weak. The audience expects speakers to draw conclusions from their presentation. They should do so, and ask confidently for a commitment because a commitment is in the audience's interest as well the speaker's.

HOW TO CONCLUDE A PRESENTATION

There are a number of techniques for bringing a presentation to an end. Whichever one is chosen the speaker should, in every presentation, be concentrating on the needs of their audience when they wind up, so that the audience's minds are being focused on their needs, their objectives, rather than the speaker's.

Audiences tend to remember best what they have heard *at the beginning* and *at the end* of a presentation. From a structural point of view, it helps to:

(1) Refer back to the theme of your presentation expressed in terms of audience needs, for example, 'Ladies and gentlemen, let me return to the objective I set out to achieve for you.'
(2) Summarize the main points of your presentation, for example, 'At the outset, I aimed to deal with these points and how they relate to you.'
(3) State your plan of action for your audience in a precise, orderly fashion. Don't leave the audience with a bundle of generalities. For example: 'First we must decide what we want to do about the information we now have about this medical research. Then we can set specific objectives for the whole programme. Secondly we must decide how we are going to reach those objectives. Finally, we must agree on a formula which we can use to judge how well or badly we are reaching our objective – in short we must have a control system.'

Asking for action

When asking your audience to commit themselves to taking action, use one or more of the 'closing techniques'. Whichever you decide to use, *write down and memorize the exact words with which you are going to close*. This will enable you to look at your audience, to look confident and to end on a high note.

Closing techniques

Here are examples of the six techniques most frequently used to close a presentation.

1 Closing by summarizing, for example:

'You want surgical equipment that will achieve three things for you:

(a) It should be capable of reducing operating time as well as being easy to use
(b) It must be compatible with the theatre equipment you wish to retain
(c) It must be installed and ready for operational use within six weeks.

Taking these needs into consideration, your best approach would be to choose our equipment because it meets all the criteria.'

2 Closing on a story coupled with a request, for example:

> John Bright arrived at the scene of a street accident. Seeing the serious injuries of the victim, he took off his hat, put £10 into it and went round the onlookers saying, 'I am £10 sorry, how sorry are you?'

3 Closing with a literary quotation coupled with a request:
Writers frequently capture in superb language what we would want to express and there is no shame in quoting a piece of poetry or prose written by someone else, if it is apt and will leave an indelible memory in the minds of your audience: 'Tis not in mortals to command success, But we'll do more, Sempronius; we'll deserve it.' 'When one door closes, another opens, but we often look so long and regretfully upon the closed door, we do not see the ones which open for us.'

4 Closing by asking directly for action
This technique is best used when you are reasonably certain that the audience is going to give an unequivocal 'yes'. Because if you don't get it, the only other answer is 'no'. For example:

> Can we take it that you are all fully in support of this proposal and want to go ahead? Can I have a show of hands to indicate your approval please?

5 Closing with an appeal for immediate action, for example:

> As you know, your pharmaceutical competitors have been ominously quiet during the past eighteen months. But it is believed that they are now about to launch a new product in this therapeutic area at any time now. To protect your share of this market, I suggest that you speed up the administration work on your field trials so as to make things as difficult as possible for them.

6 Closing by offering a choice of alternatives
This is frequently a much more effective way of ensuring action than presenting an ultimatum, for example:

> From what you have seen, your problem can be solved either by a comprehensive plan which would be ready in three months or by a gradual process starting now and spread over six months. Which would you prefer – to start now or wait for the comprehensive plan to be ready?

Conclusion

The end phase of a presentation should flow, and appear a logical development of what you have previously said at the beginning, in the middle and in your summary. Asking for a commitment from an audience does not mean that a favourable one will always be as easy to achieve, as for example, when the chairperson or Master of Ceremonies asks an audience to stand for the national anthem.

Action will result only if the rest of your talk, speech or presentation has been audience-oriented throughout, from your opening words to your closing sentence. A rough check of its audience orientation is to note the number of times 'you' and 'your' are used compared with 'I', 'me', 'we' and 'our'. Once again, to ensure that your message reaches its target – your audience – remember the old adage:

> Tell 'em what you are going to tell 'em. Then tell 'em. Then tell 'em what you told 'em.

To complete the planning, Figure 5.1 shows a chart that links together all the main elements involved.

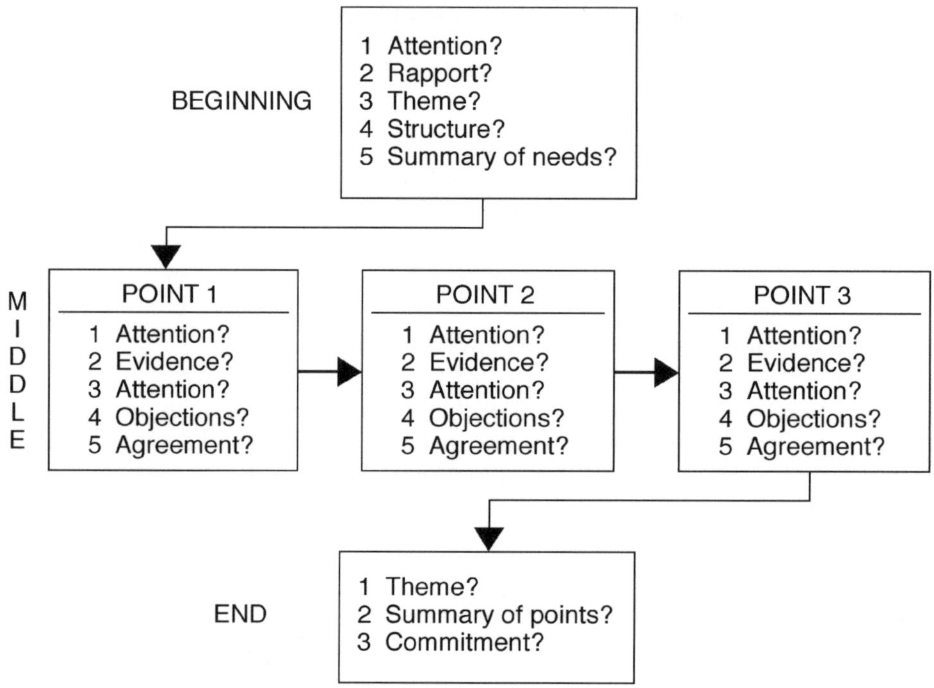

Figure 5.1 Presentation planning chart

CHAPTER

6 *Using Visual Aids*

Visual aids skilfully deployed in a talk or presentation can often convey an idea or piece of information more effectively than a verbal description. But, while it is true that *a picture is worth a thousand words*, you must be clear in your own mind *what* visual image you want to leave in the minds of your audience at the end of your presentation. If you are not careful, your visual aids can give a totally wrong impression.

All too often speakers visualize what is better said and say what would be more effective visualized. So how do you decide what to do about visualizing your presentation? Once you have compiled the information for your presentation, ask yourself *four* questions:

(1) *Do I need visual aids at all?*
(2) *What visual aids are appropriate?*
(3) *What important points of my presentation would be better understood by means of a visual aid/ picture than by words?*
(4) *What picture or memory do I want the audience to take away?*

The answer to the last question helps you to choose the one key visual aid for your presentation. It could be the only one you need.

Characteristics of good visual aids

When planning your use of visual aids, consider the following points.

(1) *They must be readable (or understandable) from the furthest point in the room.*
Always, before your presentation, go to the point furthest in the room from the screen on which your visual aid will be shown, or the table on which a model or exhibit is to appear, and check that what you plan to visualize can be both *seen* and *understood* by everyone attending. Make allowances for the fact that, from the back of a chair, the average adult's body when sitting extends to about another two feet above it. If your visual aid is not of uniform lettering/word size, then you must carry out this check for each visual aid you propose showing.
(2) *They must produce an immediate impact.*
A visual aid is used to gain or regain attention that may have wandered during your introductory remarks. Sketches, pictures or diagrams have more impact than words or figures.
(3) *Use as few words as possible.*
Avoid too much detail on one visual aid. Keep information to *four* lines at most. Letters

when projected should be at least three inches high. Keep numerical information to a minimum: people do not remember figures easily. Many presenters who show figures show too many and they are nearly always too small for anyone to see, still less understand!

(4) *They must be easy to read and simple to understand.*
Use short words, none at an angle or upside down.

Types of visual aid

There are a large number of different methods of visualizing a talk or presentation. They include:

blackboards
whiteboards
magnetic boards
flipcharts
overhead projectors
16mm film projectors
35mm slide projectors
models
three-dimensional drawings.

It is important to use the most appropriate visual aid taking into account:

physical environment in which presentation is made
size of audience
type of presentation
the message you want to deliver.

Approaches to different media

The four most common methods of visual display used in presentations are:

computer-based presentation packages
35mm slides
flip charts
overhead projectors.

The advantages and disadvantages of each are set out on the following pages.

COMPUTER-BASED PRESENTATION PACKAGES

More and more presenters now use their personal computer (pc) to prepare presentations, and for many speakers they have replaced the traditional overhead projector. There are a variety of computer-based presentation packages that enable you to prepare a screen or a series of screens containing text, information, multilingual tool support, graphics, photographs, animation and sound. You can prepare your presentation at your home or office, then take it to the venue where you are going to speak. Your audience can gather round your pc screen. If the audience is a big one, you will need a special piece of equipment to project the computer images you have created on to a large screen.

Preparing a presentation programme requires appropriate software, such as Microsoft PowerPoint, Lotus Freelance Graphics or Corel Presentations. If your pc has one of these programs installed, use that rather than buying a new one. Essential hardware: a CD-ROM drive and a soundcard. For medical directors, marketing and product managers who make frequent presentations, a notebook pc is a useful alternative to a desktop pc.

Most digital cameras are designed to work for both main types of home and office computer. These cameras now include the standard software kit needed to connect it via a dedicated lead so that you can download images from the camera to your computer. Some digital cameras are compatible with only one type of pc, so check this point before embarking on what can be a considerable expense.

ADVANTAGES	DISADVANTAGES
Professional image	Danger of making too many visuals
Remote control by speaker	No last-minute changes possible
Combines visual, sound	You are dependent on technology working!
and animation if required	Danger of cramming too much on to each visual

35 mm SLIDES

Should you decide to use 35mm slides to illustrate your presentation, there are some elementary preparations you need to carry out.

(a) *Check that the slides are*:
   (i) numbered with a small finger spot sticker in one corner and that the same number is recorded against the place in your presentation script where you plan to show each one;
   (ii) the right way up and the right way round in the carousel or magazine. Check this *before* and *after* loading and for every slide;
   (iii) separated by a blank slide between each one you plan to show your audience. This is a wise precaution. Many speakers accidentally press the operator button twice when they have finished showing a particular slide. Unless there is a gap between it and the next one, up comes the next slide before they are ready to show it. This can impair its message and impact, or damage the speaker's control of the presentation.

(b) *Check the projector*
   (i) Make sure you know how to operate the slide projector.
   (ii) Check the machine's power supply.
   (iii) Cheek that you have a spare bulb in case the present one fails.
   (iv) Check that there is sufficient length of cable to reach the power plug, from where the machine should stand, to achieve satisfactory projection.
   (v) Check how to load the machine with your slides. Then have a rehearsal, going

through the slides and checking them against the order and numbers marked on your script.

(vi) Remember that if you are in an hotel, the staff are rarely of much technical help.

(vii) Before you start showing your slides, make sure that the machine's lens projects the slides clearly and not out of focus.

Fix a piece of sticky tape where each of the four legs of the slide projector should stand, in case the machine has to be moved before your presentation takes place. If it has been moved, once again check that the projector projects the slides correctly on to the screen.

(c) *Check the meeting room*

(i) Check that the room is laid out as you require, for example in theatre, classroom or horseshoe style. Everyone seated should be able to see your slides clearly.

(ii) Check that the room can be blacked out and that the curtains/blinds are working and, when drawn, do not leave a patch of daylight/sun shining on to the screen which could make it difficult for the audience to see the slides projected on to it.

(iii) Check that there are no seats from which the audience's vision of your slides will be masked, for example by a beam, arch or pillar.

(iv) Find out where the electrical sockets and light switches are located.

(v) Check whether there is a microphone, that it is working and how to adjust the level of sound.

(vi) Check where the lectern or table from which you are to make your presentation is placed to ensure that, when you are showing your slides, you do not mask them by standing in front of either the screen or the beam projecting them on to it.

(vii) Note where the entrance doors are. If possible avoid having a room set out so that the entrance doors are in front of the audience. If this happens late arrivals will distract attention from your presentation.

(viii) Make sure that, if your meeting is being held in an hotel, the staff have a copy of your meeting agenda and the timing of your slide presentation. You do not want it upset by any interruptions, least of all by the clatter of a trolley laden with coffee cups being wheeled into the room.

(d) *Murphy's Law check – 'if anything can go wrong, it will'*

Things can go wrong in conference centres and public meeting venues such as hotels. Here are a few extra checks that personal experience suggests are worth making. If your meeting/presentation is held in an hotel or conference centre:

(i) Check that the receptionists/management have reserved the room for the time you require for your presentation, including sufficient time before the meeting is due to start to allow you to prepare. On one occasion, I was the guest speaker at an all-day meeting of a company at an hotel. When I rose to speak, a team of waiters arrived to set up the room for an evening banquet.

(ii) Check that there are no building repairs going on – a pneumatic drill is a powerful competitor to even the most articulate speaker.

(iii) Check that your meeting room is sound-proofed against any noise that might be made in an adjoining meeting room. I was chairing a meeting once of senior management for a one-day planning session. As I began my introduction to the day's activities, my voice was completely drowned as 300 hearty salespeople in the next room began singing their company's song to the tune of 'Its a long way to Tipperary'!

ADVANTAGES	DISADVANTAGES
Professional impact	Require darkened room
Remote control operation by speaker	Require suitable white wall or screen for projection
Easily carried	Require long throw for picture unless equipment has wide-angle lens
Neat/compact projector	No masking of any part of projector slide possible – lecture control minimized
	No last-minute changes possible
	Difficult for audience to take notes in dark
	Can be costly to make

FLIP CHARTS

If you propose visualizing your presentation by using flip charts that you plan to prepare yourself, here are some tips.

(a) Do not show you first visual until needed. So the top flip chart should either be blank or have only the title of your talk on it.
(b) Have each flip chart covered by at least two thicknesses of plain flip chart paper and interleave the same amount between each of the others you plan to show your audiences. This avoids the danger of the outline of your next visual being seen through the flimsy newsprint type flip chart paper usually supplied and thus competing for the attention of the audience, which you want to revert to you after the last flip chart shown has done its job and been turned over.
(c) When you want to illustrate your talk with a diagram, it is not always possible to judge that the result will fit the size of the paper. So in advance, using an HB pencil, trace in very faintly the outline of the diagram you want to show. Then you can confidently draw boldly over your own tracing with a felt-tipped pen knowing that the result will not run off the page. Oh, and you will also probably win some brownie points for professionalism from your audience.
(d) Dependent on whether you write with your left or right hand, you should place the flip chart stand to your left or right side so that when either turning over the charts or writing on them, you do not mask them from your audience.
(e) If you are not very good at making flip chart visual aids, but want to use this medium, your local art college is a good and reasonably priced source of talent.

ADVANTAGES	DISADVANTAGES
No mechanical power resources required	Expensive if professionally made for large audiences
Can be table-top size for small audiences or free-standing larger type for big audiences	Portability problems if large

Can be modified instantly	Masking difficult
Can be prepared free-hand on the spot	Limited audience size
Can be used in daylight so audience can take notes	Not as professional as a 35mm presentation
	For larger flip charts, easel or hanging facilities needed

OVERHEAD PROJECTORS

The projector continues to be one of the most popular means of visualizing parts of a presentation, probably because the presenter can face the audience whilst illustrating a talk.

ADVANTAGES	DISADVANTAGES
Can be used in a lit room or in daylight conditions	Unless carefully arranged audience's view can easily be obstructed
Projects a large picture from a short distance	Professionally made colour visuals are expensive
Visual aids can be masked or overlaid	Visual aids do not appear as professional as 35mm slides
Freehand visuals can be prepared or written during a talk	Require white wall or screen
Alterations can be made at the last minute	Unless screen is angled, visual aids projected have a tapered appearance.
	Heat from projector lamp can cause transparencies' lettering to become smudged and unreadable.

Size

Provided the screen is well illuminated, the letters on an overhead projector transparency of one inch will suffice for an audience sitting at a maximum of 25 feet away. An audience's viewing angle should be carefully controlled, for example the wings should not be too oblique; the screen should not be too high.

Allow at least 12 square inches, and preferably 24 square inches, per head of audience in terms of viewing area so that everyone can view with a degree of comfort. You can calculate the maximum distance that an audience should be from the screen by following the simple formula:

$$\frac{\text{Height of screen in inches} \times 10}{12} = \text{distance in feet from screen}$$

REMEMBER it is better to make sure by preparation that everyone in your audience is going to be able to see and understand your visuals than to ask the audience once you have started whether everyone can see. By then, it is too late.

Colour

The most common contrasting colours in use are white (chalk) on black (board) and black (felt pen) on white (board or flip chart). If you have a choice, it is generally believed that the white on black is marginally easier to read than black on white.

Although colours can introduce variety to a presentation, beware of pastel shades, for example pink on a red background or yellow on white.

Legibility

One of the best ways to lose audience attention during an audio-visual presentation is to project material that is not legible to the entire audience. If the speaker has to say, 'You probably can't read this from where you are sitting, so I'll read it to you,' the presentation has not been adequately planned.

Once the objectives and the strategy of a talk have been decided, consideration can be given to the size of the anticipated audience, as well as to any unusual features of projection facilities. Only *then* should the artwork be designed. If a presentation is to be successful, original art must be prepared *with the people in the rear seats in mind!*

Experience shows that:

(1) Artwork can be planned and executed to permit the visuals to be legible when projected.
(2) There are advantages in establishing uniform sizes for artwork and making these sizes standard.
(3) Although the letter height can usually be a *minimum* of 1/50 the height of the information area, a larger letter height (1/25 or more) is strongly recommended.

LEGIBILITY REQUIREMENTS

To be legible, lines, letters and symbols should contrast adequately with the background; there must be separation of tones, and the colours selected should be strong and attractive. Tonal contrast is particularly important when preparing artwork for television where the television screen may display the coloured artwork in a black-and-white mode.

Letters and symbols should be bold and simple, with no small openings that will tend to fill in when projected. All elements such as lines, letters, symbols, and figures require a size big enough to be seen easily by everyone in the audience. Therefore these elements have to be at least a certain minimum size on the screen, the actual size depending on the height of the artwork area in relation to its distance from the farthest viewer.

In typical viewing situations the maximum viewing distance should be about eight times the height of the projected image. To put it another way, if the projected material is legible for the farthest viewer, it will be legible for all other members of the audience. This maximum viewing distance (expressed as 8H) can be used to determine the minimum size of significant details in the material projected.

TESTING EXISTING MATERIAL FOR LEGIBILITY

When material that was not designed for projection (printed graphs, charts, and so on) is to be converted to a projected visual, remember that although contrast and viewing distance may change the requirements for legibility remain the same.

Note that 8H viewing is a generally accepted standard. If the letter size suggested for 8H viewing is doubled, the projected image will be legible from twice the distance or 16H. The 8H concept also assumes average or slightly lower than average eyesight on the part of the viewer. For 8H viewing, legibility can be judged by an average viewer by looking at material to be copied from a distance eight times its height. For example, consider a printed table that is to be photographed for projection. If the table is 3.15 inches (88 mm) high, it should be viewed from eight times that height (28 inches or 0.7 m) to see if it is readable. If it is, the type size will be suitable for copying and projection.

The same principle applies to larger work. A wall chart or map 4 feet (1.2 m) high requires legibility at a distance of 32 feet (9.6m) if it is to be acceptable as a projected image for 8H viewing (4 feet × 8H = 32 feet). If the material is not legible at the test distance, it should be either redrawn or discarded.

Subject *content* as well as image size affects legibility. If the work you are photographing is complex, reduce the information to the essential elements, limit the text, and enlarge the letter size. Rearranging the information can help define the point you are making for the audience.

It is a mistake to believe that enlarging the physical dimensions of a transparency improves legibility at practical viewing distances. *Transparency size* is not a determining factor; it is the size of the detail on the *screen* that is significant.

If the letters are to be legible at an 8H viewing distance of 32 feet, a projected image of 1 inch (25 mm) high on the screen is required whether projection is from a 2 × 2 inch (50 × 50 mm) slide or a 10 inch (250 mm) wide transparency on an overhead projector, regardless of the overall projected image size.

STANDARDIZATION

We have indicated that a minimum size for lettering has been established to meet legibility requirements. But legibility is not the only requirement for effective communication; flexibility, so as to allow emphasis and pleasing design, is also important. It will therefore be wise to standardize on at least three letter sizes to provide proper treatment and a variety of titles – primary, secondary and tertiary. The use of more than three sizes, all larger than the recommended minimum, allows even greater variety in artistic freedom. Of course, standardizing letter sizes is practical only when the format and overall size of the artwork are also standardized.

The cost per hour of skilled professionals, such as artists and photographers, far outweighs the cost of materials. Thus the largest savings to be realized in preparing a visual presentation lie in reducing the time required to complete it. Standardizing format and size of artwork will yield the greatest cost reduction. There are other benefits to standardization. One is that the artist can work with a few standard, readily available pens, brushes, guides and type sizes. A feel for the size of lettering and artwork elements that will produce legibility can quickly be developed. Therefore, standard-size artwork becomes easier and faster to prepare than the alternative – an assortment of various sizes and shapes. Standard sizes simplify the stocking of mounting boards and paper stock. Making the artist's and photographer's job less time-consuming can increase productivity without increasing cost.

A standard field size for artwork and a specified location for the working area on the artwork can speed the photography and consequently increase the photographer's output. When working with artwork of random sizes and formats, the photographer needs to adjust the camera-to-artwork distance, the focus and the exposure only *once* for each complete *assignment* rather than once for each individual piece of artwork if the following conditions are met:

(1) The artwork is all the same size.
(2) The working area is of the same dimensions on every piece of artwork.
(3) The working area is in an identical location on each piece of art.
(4) Provision is made for placing each piece of artwork in the same position on the copy stand.

Adopting a uniform 10 × 12 inch (250 × 300 mm) artwork size offers savings in cost and time. Storage of this size requires no expensive equipment – letter-size office filing cabinets or desk drawers will serve. Artwork can be stored on edge, and segregated into categories with standard separators. The material is readily accessible, while the possibility of damage or loss is reduced.

Numbers, statistics and financial data

Numbers on visual aids, particularly when used in comparisons, lose a great deal of impact if merely left in number form. How often have you heard a speaker say of the next visual aid on the screen, 'You will probably not be able to read these figures too well, but they show a decline in sales in the third quarter.'

Charts should aim to communicate numerical information in a more digestible and easily understood form than the spoken word. There are five main chart forms: pie, bar, column, curve and dot. Your choice should be based on how best to illustrate the comparison in question.

1 NUMBERS

Look at the contrasting visuals in Figures 6.1 and 6.2. The first lists the world sales of top pharmaceutical products in 1998 in number form. You need to study the details very closely to understand the information being conveyed. Whereas Figure 6.2, which shows in pie chart form share of sales of the world's top prescription medicines for the same year, a circle gives a clear indication of the relative sizes or components that make up a whole. Remember to slice the pie into not more than five different shades or colours. In the one illustrated, the largest slice, the two-thirds share held by the USA, leaves an indelible picture in the minds of your audience – which is what a good visual aid should do.

1998 Ranking	1997 Ranking	1998 World sales £ million	Product name	Company	Country of origin
1	1	2,686	Losec	Astra	Sweden
2	2	1,784	Zocor	Merck	USA
3	3	1,568	Prozac	Lilly	USA
4	5	1,412	Norvasc	Pfizer	USA
5	41	1,167	Lipitor	Warner-Lambert	USA
6	6	1,081	Renitec	Merck	USA
7	9	1,022	Seroxat	SmithKline Beecham	USA
8	7	1,010	Zoloft	Pfizer	USA
9	8	916	Augmentin	SmithKline Beecham	UK
10	14	884	Claritin	Schering-Plough	USA
11	10	860	Ciproxin	Bayer	Germany
12	15	844	Epogen	Amgen	USA
13	40	778	Zyprexa	Lilly	USA
14	16	759	Pravachol	Bristol-Myers Squibb	USA
15	39	747	Ogastro	Abbott	USA
16	11	738	Klacid	Abbott	USA
17	4	725	Zantac	Glaxo Wellcome	UK
18	33	691	Flixonase	Glaxo Wellcome	UK
19	12	684	Voltaren	Novartis	Switzerland
20	13	657	Adalat	Bayer	Germany
21	18	642	Imigran	Glaxo Wellcome	UK
22	27	616	Risperdal	Johnson & Johnson	USA
23	31	614	Zithromax	Pfizer	USA
24	23	593	Neupogen	Amgen	USA
25	36	592	Erypo	Johnson & Johnson	USA

Sources: IMS
ONS Financial Statistics March 1999

Figure 6.1 World sales of top 25 pharmaceutical products, 1998

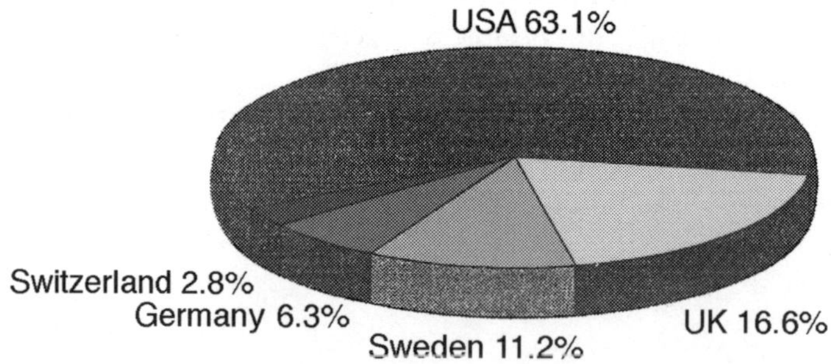

Source: IMS.

Figure 6.2 Share of sales of the world's top 25 prescription medicines, 1998

2 STATISTICS

Rows and rows of statistics are often much clearer when presented in graphic form. Indeed, for many people column or bar chart form is better still. First, observe the two sets of statistics: Figure 6.3 lists NHS expenditure per person, Figure 6.4 NHS prescriptions dispensed. For most people such statistics are difficult to comprehend or remember. Now

Year	Total NHS cost[1] £m	Total NHS cost per medicines[2] £cash	All NHS medicines[2] £m	NHS prescription medicines cost[2] per person £cash	Medicines as a % of total NHS cost
1980	11,521	222	826	38	8.9
1985	17,641	345	1,627	48	9.8
1990	28,336	554	2,533	57	8.9
1991	32,124	630	2,755	58	8.6
1992	35,796	699	3,189	64	8.9
1993	38,398	747	3,893	75	10.1
1994	40,028	784	4,227	81	10.6
1995	41,848	817	4,583	85	11.0
1996	43,418	854	5,078	91	11.7
1997	45,368	893	5,551	96	12.2
1998	48,238	946	6,056	102	12.6

Notes: 1 Includes charges to patients
2 Excluding dressings and appliances, at 1998 manufacturers' prices

Sources: United Kingdom National Accounts – Blue Book (ONS)
Prescription Pricing Authority Annual Reports
NHS Summarised Accounts
Annual Abstract of Statistics (ONS)
Compendium of Health Statistics 1999 (OHE)
Health and Personal Social Services Statistics for England
Health Statistics Wales (Welsh Office)
Scottish Health Statistics (ISD)
Statistical Report (Northern Ireland CSA)

Figure 6.3 NHS expenditure per person in the UK

Year	Prescription irems (m)	Prescription items per head	Total rescription cost[1] £m (cash)	Average NIC[2] per prescription £ (cash)	Average cost per in-patient day[3] £ (cash)	Exempt prescriptions as a % of all prescriptions
1970	306	5.5	209	0.54	8	n/a
1975	346	6.2	448	0.95	22	n/a
1980	374	6.6	1,119	2.37	48	n/a
1985	393	6.9	1,875	3.91	84	74.9
1986	397	7.0	2,031	4.21	95	76.4
1987	414	7.3	2,262	4.55	102	77.3
1988	428	7.5	2,527	4.98	109	77.6
1989	436	7.6	2,728	5.32	121	77.8
1990	447	7.8	2,984	5.74	130	78.4
1991	468	8.1	3,343	6.15	142	80.0
1992	488	8.4	3,729	6.67	154	81.0
1993	512	8.8	4,091	7.01	162	82.1
1994	524	9.0	4,365	7.36	169	82.7
1995	545	9.2	4,711	7.67	171	83.8
1996	560	9.5	5,096	8.12	181	85.6
1997	578	9.8	5,448	8.55	192	85.4
1998	594	10.0	5,788	8.96	n/a	85.5

Notes: 1 Dispensed by chemists and appliance contractors
1 Includes medicine cost, dispensing fees, allowances etc
2 NIC = Net Ingredient Cost (before discounts)
3 Figures are for financial year ended 31 March, England only

Sources: Health Database (CIPFA), Compendium of Health Statistics 1999 (OHE)
Annual Abstract of Statistics (ONS)
Health and Personal Social Statistics for England (DoH)

Figure 6.4 NHS prescriptions dispensed in the UK

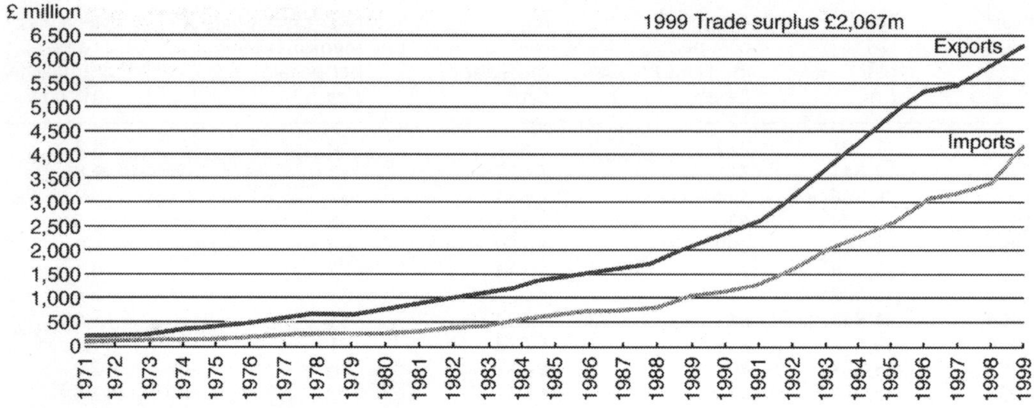

Note: 1999 figures are estimates.
Source: Business monitor MM20 (HM Customs & Excise).

Figure 6.5 UK pharmaceutical trade

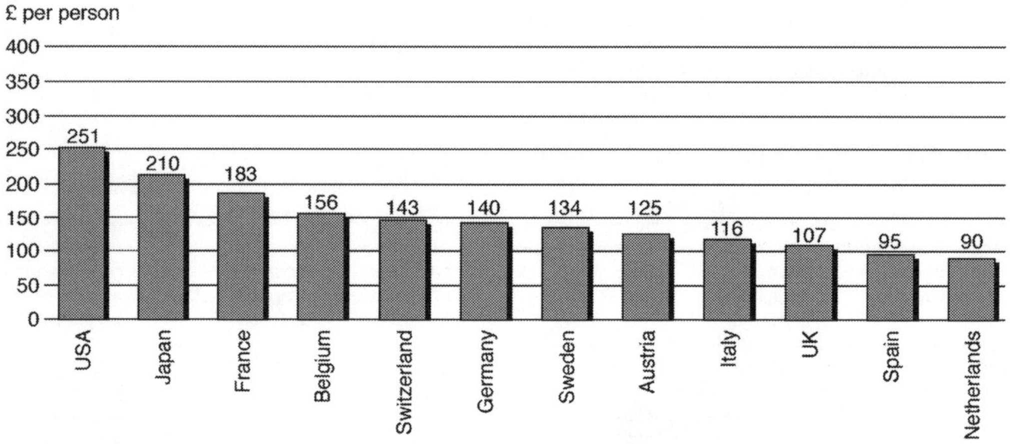

Note: Ex-manufacturers' prices.
Source: IMS World Review 1999.

Figure 6.6 Per capita sales of pharmaceuticals, 1999

look at two different visuals, Figure 6.5, a graph showing UK pharmaceutical exports and imports from 1971 to 1999, and Figure 6.6, a bar chart showing per capita sales of pharmaceuticals for twelve countries, ranked in descending order. Such a chart shows at once how sales in the USA, Europe and Japan compare. These two visuals, in graphic and bar format, communicate clearly whereas dense sets of figures set out in large quantity and tiny type will leave an audience bored or bemused.

3 FINANCIAL DATA

When a presentation involves making a case in profit and loss terms, price factors, market share, expressing NHS spending by percentage or by amount and the like, there are a variety

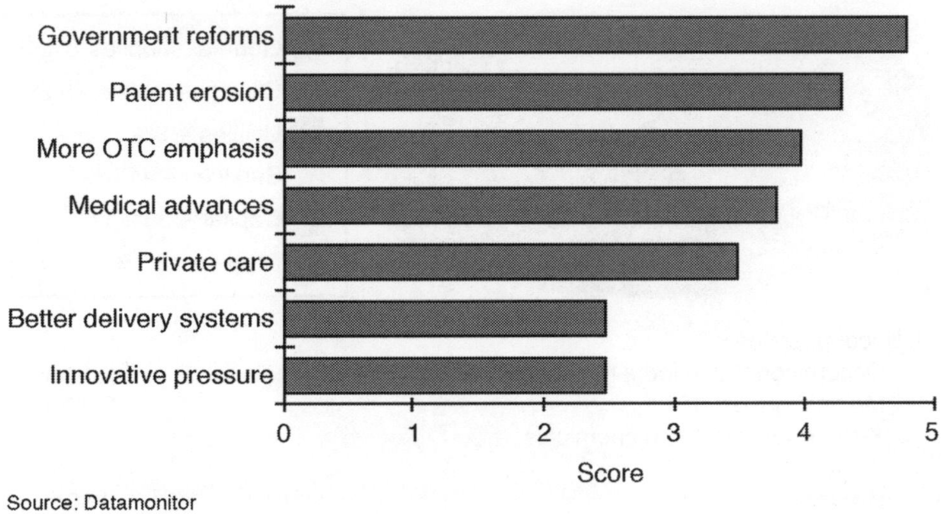

Figure 6.7 Factors influencing EU pharmaceutical prices

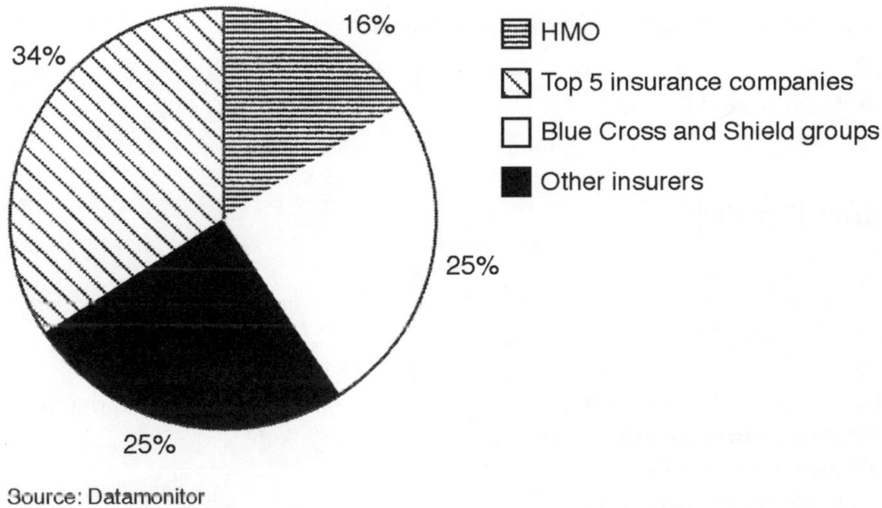

Figure 6.8 Shares of the US health insurance market

of suitable visual presentations, including horizontal bar charts and pie charts. Some examples are given in Figures 6.7, 6.8 and 6.9.

Compare the horizontal bar chart (Figure 6.7) with the vertical one shown in Figure 6.6. It is usually easier to label horizontal bar charts than vertical ones and easier, too, for your audience to read the labelling provided it is in bold lettering. In the pie chart 'NHS spend' (Figure 6.9), it would be much more effective to present the Clinical Supplies breakdown in large bold figures on a separate follow-on slide than to show it alone.

Weigh up from the point of view of the audience, wherever they are seated, what would be the 'best' way to visualize such information. Remember that what matters is not what appeals to you artistically but whether your audience will understand pictures more readily than your verbal description.

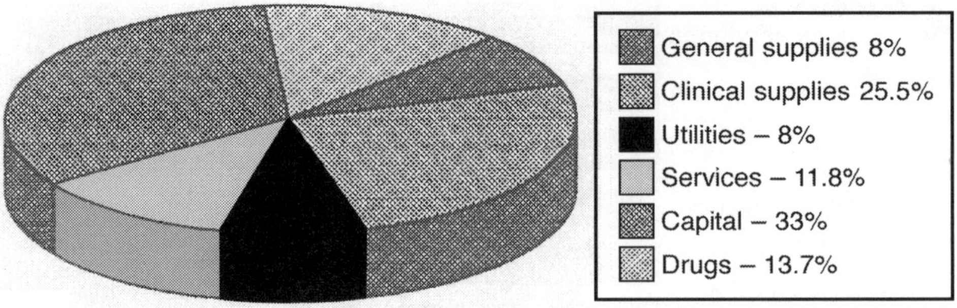

Clinical supplies	
Occupational and industrial therapy equipment and material	£ 14m
Medical gases	£ 23m
X-ray equipment and chemicals	£ 27m
X-ray film	£ 43m
Dressings	£ 86m
Artificial limbs and hard-core equipment	£ 91m
Laboratory equipment instruments and materials	£152m
Patients appliances	£162m
Medical and surgical equipment purchases	£702m

Figure 6.9 NHS spend

Dos and Don'ts

DOS

keep visual aids simple
check that visual aids are readable from furthest point in room
use as few words as possible – choose key words and phrases that are memorable
design visual aids to relate directly to the words being spoken
use visual aids to complement a presentation
use correct colours in visual aids
use charts rather than figures
rehearse using your visual aids
ask 'Do I need any visuals at all?'

DON'TS

use a visual aid if you have any doubts about it
visualize what can be just as well said
use visual 'verbals'
take your audience through your slide presentation too quickly; some may want to make notes from them
be afraid of silence
compete with your visuals – let them speak for themselves

switch your overhead projector on and off without reason
mask your visual aids by standing in your audience's line of sight
talk to your visual aids.

Summary

Here, to conclude, are some rules for using visual aids to best effect:

(1) Identify the parts of your presentation where a visual aid will help communicate your ideas better than just relying on words.
(2) Select your visual aids – too few rather than too many.
(3) Ask yourself Do they say what I want them to say?
Are they as simple as possible?
Could they be misunderstood?
Will they distract the audience?
Could the machinery break down?
Will they work?
(4) Prepare your visual aids thoroughly, no matter how simple. Very basic actions, such as pencilling in outlines or a checklist of headings on a flip chart before presentation, can turn an ordinary talk into a most professional one.
(5) Keep them hidden until required.
(6) Introduce your visual aid before showing it.
(7) Let your visual aids speak for themselves. Give the audience time to read or study them before proceeding.
(8) Any comments you make while showing a visual aid should be limited to brief, concise explanations of the content.
(9) Check to ensure that everyone understands before moving on. There is no point in developing the subject further if some in your audience are confused.
(10) When your visual aid has done its job, remove it from sight; otherwise it will compete for your attention on the next point you want to make.

Visual aids can help you to improve the chances of your presentation's success. After all, if you speak, you are only using one of the five senses through which we send and receive messages. The fewer you use, the greater the possibility of your message not being completely understood. Whereas if you show a picture to your audience you are appealing to a second means by which we send messages – sight.

Visual aids well designed and deployed can attract attention that has wandered, add a picture to your words, hold or regain attention, simplify what was complex, create a memorable vision and summarize.

CHAPTER 7 Video Conferencing

It is an accepted part of business life, especially in multinational pharmaceutical companies, for executives to travel great distances to attend meetings, including medical, financial, manufacturing, sales and management conferences, and board meetings. Now everyone must weigh up the consequences of such travel. Why?

First, after the attacks in America on 11 September 2001, in Bali and Mombasa in 2002, and the Iraq War and SARS outbreak in 2003, people are less willing to fly. The pharmaceutical industry is global in its compass. Executives, as a matter of course and often daily, travel within their own country and to other countries, attending staff meetings, medical symposia and international conferences. When I directed my company's healthcare division, in one year alone I visited thirty-six countries.

Secondly, travel now involves huge expenditures for companies, not just in air, rail and road travel, but also in opportunity costs. British executives, for example, spend over ten hours travelling to attend an average of six meetings a week. The amount of time wasted on roads, in trains and on aircraft is immense.

Thirdly, technology such as video conferencing can replace more than ten per cent of the business travel undertaken. About twenty-five per cent of all miles driven in cars are work journeys. You can imagine the effect in terms of congestion and delayed meetings, not to mention the tempers of everyone concerned. In addition to all this, two-thirds of top executives suffer serious physical symptoms from business trips, quite apart from the disruption to family life caused by absences from home and week-ends cut in half by departures to airports for the start of another week's round of meetings, often on the other side of the globe.

Meetings still take place, notwithstanding the events of 11 September. But their number, and the number of those who travel by air to attend them, has diminished dramatically, which is why airlines are facing serious financial difficulties.

Although video conferencing has been used for years, until recently the investment, involving as much as £50,000 to install, plus the running costs, meant that is was considered too expensive as a method of communicating outside company boardrooms. As with all developing technologies, both the capital and the running costs have fallen to the point where video conferencing is now an affordable, everyday tool of communication for businesses and staff at all levels. And cheaper systems, costing as little as £1,500, can interact with the much more expensive ones like those to be found in larger pharmaceutical companies.

Video conferencing is now used by most British, European and American pharmaceutical companies, including: Abbott, AstraZeneca, Aventis Pharma, Boehringer Ingelheim, GlaxoSmithKline, Merck, Novartis, Pfizer, Hoffmann-La Roche and Sanofi-Synthelabo. Medical and clinical staff can discuss the progress of drug clinical trials with their opposite numbers in places conducting similar trials as far apart as New Zealand, Argentina, California and Wilmslow, UK. Doing all this in a one-hour link-up shows the benefits for all the people concerned, who otherwise would have had to fly into, say, Rahway, USA, Frankfurt, Germany, Basel, Switzerland or Sandwich in the UK.

Technical factors

There are a number of technical details that need to be dealt with if video conferencing is to take its rightful place alongside the other methods by which pharmaceutical companies can maintain effective communication with their staff, wherever they are based.

(1) Rooms dedicated to video conferencing must be lit well and in such a way that no one seated in any part is in shadow. Shadows which appear natural to the eye can be exaggerated by the camera, as can the light-and-shade effects created by sunlight streaming through a window. Such extra lighting can and should be capable of being dimmed down when the room reverts to use as a conventional meeting room or office.
(2) Video cameras should be positioned to allow participants to look towards them so that viewers can see them exactly as they would appear if they were in the same room. Placing cameras so that they look down on participants will focus on the tops their heads, while cameras focusing on the side of a person will not encourage natural conversation, since facial expressions are hidden and eye contact is not possible. The best results are obtained when video cameras are unobtrusive, so that the participants can forget that they are video conferencing and concentrate on taking part in a meeting and communicating with their colleagues. To achieve this, the best position for the camera is probably above the television monitor and focused on the participants with adjustable zooming-in facilities.
(3) Tape recordings can be made of video conferences so that afterwards medical, marketing, manufacturing and other interested executives can discuss the issues with other staff not present but likely to be involved in the follow-up actions agreed on.
(4) British Telecom and other telephone service providers can install ISDN lines to connect companies for video conferencing and have been known to do so at twenty-four hours' notice.

Social skills

One of the great benefits of video conferencing is that it enables companies to involve staff, who would not normally attend meetings where travel is concerned, because they are technical or administrative people rather than medical, manufacturing, marketing, product or sales executives. For these as well as for those actively taking part in them on a regular basis, there are some basic social skills that should be remembered.

(1) *Physical appearance.* As with any formal meeting you attend, ensure that you look smart so that the person or people you are meeting on video camera receive the best impression of you and your affiliate company colleagues, especially if this is the first time you have communicated with each other. Open-necked shirts, denims or a two-day growth

of beard do not convey a business-like attitude. A slipshod physical appearance can make those at the other end of the line wonder whether your thought-processes are equally slipshod! Remember that necklaces, chains or bangles, when they catch the light, can sparkle and hence distract.

(2) *Speaking to foreign or mixed audiences.* If you are going to be speaking to colleagues whose first language is not English have another look at the guidelines set out in Chapter 2 (p. 27). Remember especially to slow down your delivery from the normal 130–150 words a minute to about 90 words a minute so that your colleagues can comprehend what you say.

(3) *Body language.* Video conferences are just as much influenced by bodily behaviour as a meeting when people are physically present.

An example of video conference facilities

Figure 7.1 shows a video conference studio located at the headquarters of one of the world's largest pharmaceutical companies. Notice that the video camera is positioned above the monitors.

In the bottom picture, a small pilot screen enables those participating to see themselves, but is not so large as to cause a distraction. Desk control panels enable the participants to zoom the camera in for a close up of whoever is talking, or to see the details of a visual aid being shown.

Conclusion

A video conference should be prepared as well as, if not better than, any other company meeting. Here is a checklist for ensuring adequate preparation:

(1) Before the meeting, the person chairing it should possess a clear understanding of the issues to be discussed and the objective for the meeting. It is his or her responsibility to ensure that the video conference room and equipment are ready, and that those attending are all clear about their role and the contribution they are expected to make. A technician should be briefed well ahead of the meeting so that he or she can contact the participants in the other locations to make sure that a satisfactory link-up can be achieved and that they will all be ready at the appointed time.

(2) Prepare a clear, concise beginning so that everyone, both at your end and at the receiving end, is clear about the purpose of the meeting; Make sure that the meeting begins with a strong opening to command attention.

(3) Involvement and motivation must be maintained in the middle when attention can sag or stray from the purpose of the meeting. When more than one person is taking part, make sure that they make a full contribution. Likewise involve other senses through visual aids which can be clearly seen, read and understood by the listening/viewing participants. Use questions to check understanding.

(4) A video conference should end with the chairperson pulling together the threads of the discussion and summarizing the actions that need to be taken, by whom and when.

Figure 7.1 Video conferencing studio. Photographs reproduced by kind permission of Carillion Communications Ltd.

*Audio-Visual Projection**

No matter how skilfully a film, filmstrip, or slide presentation is prepared, its value as a communication tool will be lessened if it is projected poorly. Some part of the 'message' will be lost unless projected images are sharp, bright, and large enough to be seen clearly and easily by each member of the audience. And when sound accompanies the projected images, it must be distinct and loud enough to support the message effectively.

These notes will help you select and prepare a suitable room in which to seat your audience and project your programme. They will also help you choose appropriate equipment and arrange it most effectively in the room. We suggest that you approach any audio-visual projection situation by breaking it down into the following steps:

(1) Choose a room with adequate facilities.
(2) Select a seating plan and screen type.
(3) Determine the screen size and location.
(4) Choose suitable locations for loudspeakers.
(5) Select the projector location and lens focal length.
(6) Determine the required image brightness.
(7) Select a projector-lens-lamp combination to meet the brightness requirements.

Although these steps are set out here in logical order, in practice it is often necessary to compensate for less-than-ideal equipment and facilities by backtracking and modifying earlier decisions. The suggestions and data provided in this appendix will help you stage an effective audio-visual presentation in most facilities.

Room facilities

When the ideal room is not available, modifying existing conditions will improve a marginally suitable room. The room should:

(1) *be large enough* for the greatest number of viewers expected. Large auditoriums or meeting rooms with folding chairs need 5 to 6 square feet (0.5m^2 approximately) of floor space per person within the good viewing area. Conference rooms or classrooms with fixed seating require about twice that area – 10 to 12 square feet (about 1 m^2) per viewer within the good viewing area.

(2) *permit suitable light control.* For daytime projection in rooms with outside windows, ordinary window shades are usually suitable for images with contrasting colours, no greys or shadows, and no detail in the dark areas. Opaque shades (two thicknesses of ordinary shades), ordinary venetian blinds or curtains will suffice for images in which there is some shade detail. In most cases, however, complete darkening is needed for projected images that have good shadow detail and colour. You can use temporary opaque curtains, portable blackboards, or other makeshift devices to block light from the screen when the room cannot be adequately darkened. In some cases a permanent shadow box around the screen may be desirable. Special light control may not be needed for the extra-bright images obtainable with small screens (found in study carrels) or with high-gain screens.

(3) *provide needed illumination.* Light sources that provide low-level room illumination during projection (not directly onto the screen) allow note taking and help maintain a social atmosphere. However, during projection the screen image highlights should be brighter than any other surface within the viewers' field of view.

(4) *provide adequate electrical control.* Preferably, room lights should be controllable from a point near the projector or the speaker's stand. Or arrangements can be made for someone to turn the lights off as soon as an image appears on the screen. The electrical outlet for the projector must remain live when the room lights are turned off.

(5) *provide good ventilation.* The ventilation should be independent of the room-darkening devices. If smoking is permitted (preferably it should not be), a generous supply of fresh air will be needed.

(6) *be acoustically good.* Most rooms are satisfactory. Check reverberation by a smart clap of the hands. A sharp, ringing echo indicates too much reverberation for good intelligibility. An overly live room will reverberate less when filled with people. Any loud noises or clearly intelligible speech emanating from outside the room should be controlled or eliminated. Low-level background noise does little or no harm.

(7) *have a sufficiently high ceiling.* The ceiling should be high enough to allow putting a screen image up where all members of the audience can see it. A stepped or sloping floor would be ideal. (See the section on screen and image size, pp. 77–79.)

Seating plan and screen type

All screen and image dimensions mentioned in this appendix apply to picture aspect ratios commonly used in audio-visual materials. These range from squares to rectangles with the long dimension no more than 1.5 times the shorter.

When using multiple images, consider each image or image area individually, whether the images are on the same screen or different screens. That is, the requirements for size, brightness, and legibility will usually be the same for each image as if there were only that one image.

Practicality may, however, dictate a compromise. In spite of the requirements for optimum quality of each image, room size and seating arrangement may limit the size and type of screen.

FRONT-PROJECTION SCREENS

The screen is often the weakest link in a projection chain. A projection screen interrupts the light falling on it from the projector (or other sources) and reflects it to the viewers' eyes. The efficiency with which it does this affects image brightness, evenness of image brightness, colour saturation and contrast of the image. Following are brief discussions of some screen types and their qualities. Figure A.1 illustrates how room shape and screen position determine good viewing and thus good seating plans.

Many *high-gain aluminium screens* have a surface several times brighter than that on regular screens available. Because of their surface characteristics, these screens can be used in a normally lighted room. When properly positioned, this type of screen rejects room light by reflecting it away from the viewers, thereby retaining full contrast and colour saturation in the viewing area.

Matte screens diffuse light evenly in all directions. Images on matte screens appear almost equally bright at any viewing angle. To avoid distortion because of viewing angle, however, viewers should be no more than about 30 degrees to the side of the projection axis and no closer to the screen than twice the image height (2H).

Most matte screens are about 85 per cent efficient. That is, an illumination level of 10 footcandles (1081 ×) on the screen provides a screen surface brightness of 8.5 footlamberts ($29 cd/m^2$), regardless of the viewing angle or the angle at which light strikes the screen.

Lenticular screens have regular patterns of stripes, ribs, rectangles or diamond-shaped areas. The pattern is too small to see at viewing distances for which the screen is designed. The screen surface may appear to be enamelled, pearlescent, granular metal or smooth metal; it may or may not have a coating over the reflective surface.

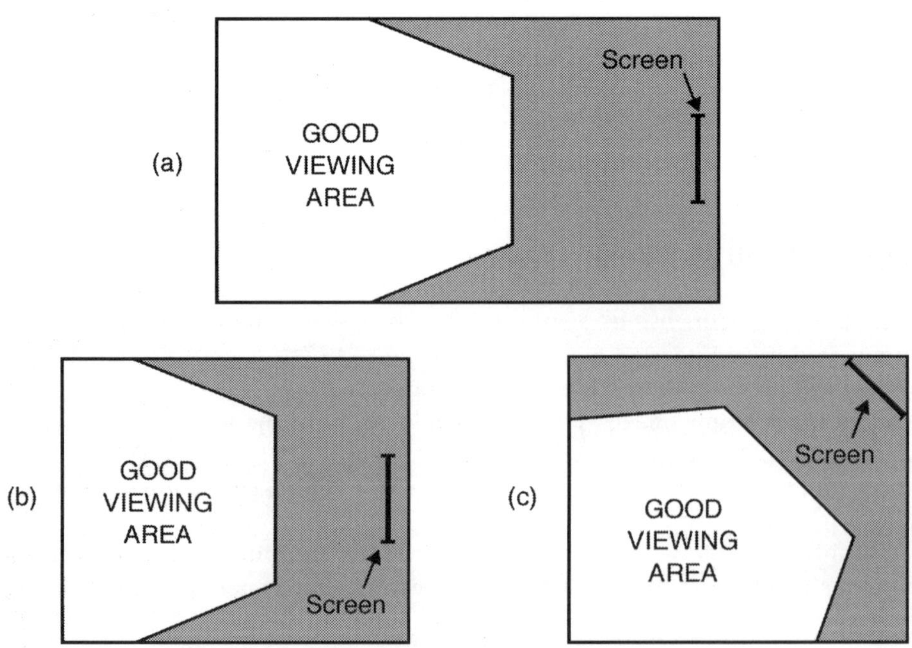

Figure A.1 Room shape and screen position

By controlling the shape of the reflecting surfaces, the screen can evenly reflect nearly all the light from the projector over a fan-shaped area 70 degrees wide and 20 degrees deep. People seated farther to the sides of the screen than the 70-degree angle or above or below 20-degree angle will see no image – no image-forming light is wasted outside the viewing area. It is also possible to produce a surface that will reject light from outside the viewing area that might otherwise reduce the contrast of the projected image.

Many lenticular screens provide an image three or four times as bright as a matte screen. Other lenticular screens may provide wider viewing angles with less gain or narrower viewing angles with more gain.

Because the characteristics of lenticular screens vary it may be difficult to select the best screen for a particular application. However, when the correct screen is used instead of the wrong one, the improvement in results will normally more than justify the effort that went into selecting the screen.

Beaded screens are useful in long, narrow rooms or other locations where most viewers are near the projector beam. Such screens have white surfaces with embedded or attached small, clear glass beads. Most of the light reaching the beads is reflected back towards its source. Thus a beaded screen provides a very bright image for viewers seated near the projector beam. At about 25 degrees from the projector beam, the image brightness on a beaded screen will be about the same as on a matte screen. Beyond this angle, it will be less bright than on a matte screen.

Since non-image or stray light is also reflected back in the general direction from which it comes, stray light falling on a beaded screen from a viewer's position at the side of a room can be a serious problem.

Screen and image size

Screen size should be such that it will permit the back row of viewers to be eight times the image height (8H) from the screen, with the following exceptions:

(1) Certain materials, including many teaching films, are designed with titles and important picture elements bold enough to permit satisfactory viewing at distances of 11 to 13 times the image height. If this is true for materials to be projected, the projector can be moved closer to the screen to give a smaller and brighter image. Moving the projector closer to change the back row from 8H to 11H will approximately double the image brightness and allow the front row to be moved a little nearer the screen.

(2) In some situations, materials that limit maximum viewing distance to less than 8H are commonly used. Typewritten material projected with an opaque projector is an example. For showing the full area of an 8.5 × 11 in (216 × 279 mm) page, pica calls for a maximum viewing distance of 4H. Figure A.2 illustrates a common arrangement used for projection rooms. The screen location was selected to cover the maximum seating area for the room. Figure A.3 shows appropriate spacing for three typical seating arrangements.

For optimum legibility, members of the audience should be seated within the specified angles for the screen material being used and should not be seated closer to the screen than twice or farther than eight times the height of the projected image. Minimum image height

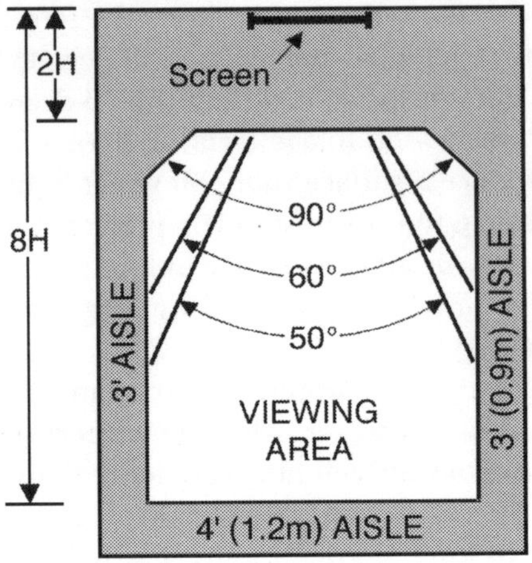

Figure A.2 A common arrangement for projection rooms

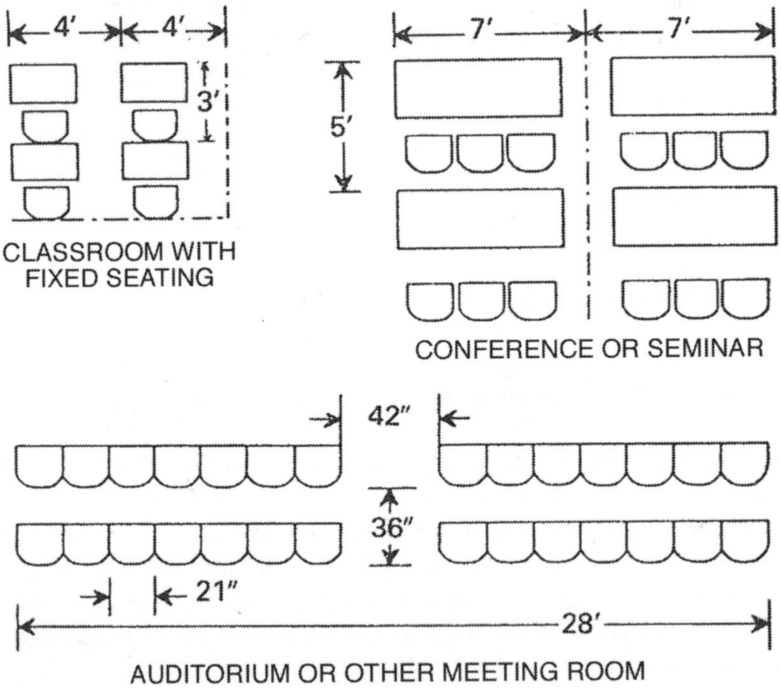

Figure A.3 Spacing for typical seating arrangements

for legibility can be determined by dividing by eight the distance from the screen to the rear of the back row of seats. For visual effect, it is sometimes desirable to project an image somewhat larger than legibility standards specify. To avoid obstruction of the screen image by the seated audience, the ceiling height should permit the bottom edge of the image to be located at least 4 feet (1.2 m) above the floor.

The maximum viewing area depends on the material used for the screen. Most matte and some lenticular front-projection screen materials can provide good brightness levels for viewing areas up to 90 degrees wide. Beaded front-screen projection materials and commonly used rear-projection screen materials can give good brightness in a viewing area up to 50 degrees, the recommended maximum viewing angle for beaded screens.

If vertical or square pictures are to be shown, a square screen is preferable. If only horizontal pictures are to be shown, either a horizontal screen with proportions of about 3:4 or a square screen masked or opened part-way is suitable.

Two-image presentations

Increasingly, programmes involving two or more simultaneous images are being used to present information. A two-image format offers special opportunities – two side by side images can show comparison of two different products, before and after conditions, or the progress of two different but related processes.

A two-image presentation can help you show details in perspective by allowing you to put a close-up view on the screen beside an overall reference view, or a photograph beside a schematic diagram. The considerations discussed here refer primarily to presentations intended to teach, train or inform. Different considerations will apply when the goal is to entertain or to motivate, and the desired effect may be an impression or an emotion rather than the assimilation of all the information and detail in each image. Essentially the same advantages and requirements apply to all three or more images side by side.

A two-image format will need a little extra forethought and may require some different equipment, particularly screens. Some slight modification of projection technique is necessary. Figure A.4 shows two screens, the same size as the screen in Figure A.1, placed side by side in a flat plane along one wall or room. The satisfactory viewing area is shown for each screen, corresponding to that shown for a single screen in Figure A.1. Obviously it is only in the unshaded area that viewing is satisfactory for both screens – and that area is rather restricted. Figure A.5 demonstrates the effect of angling the screens slightly, so that the axis of each projection beam originates from the centre of the back wall. The good viewing area for *both* screens has increased a great deal until it approximates to the good viewing area for one screen shown in Figure A.1. It makes little difference whether you use two flat screens or a single curved screen. Usually a curved screen should have a radius of about the same length as the maximum viewing distances or up to 1.5 times that distance.

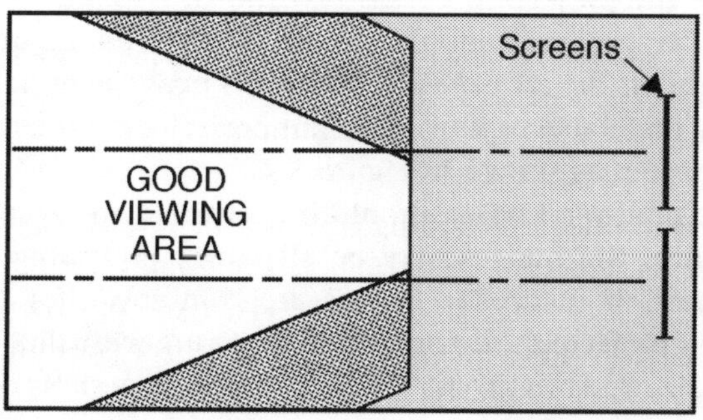

Figure A.4 Two screens side by side

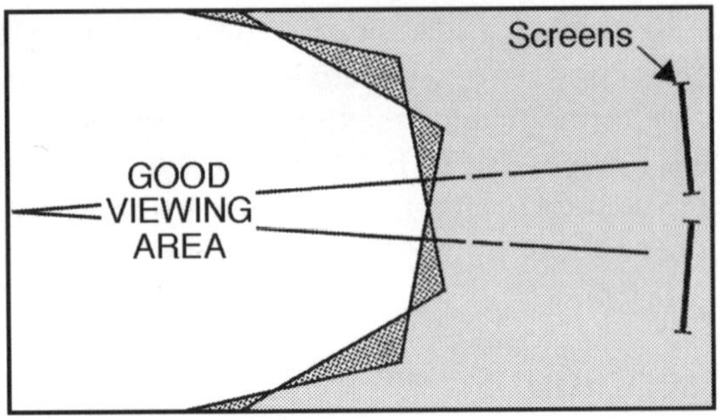

Figure A.5 Two screens slightly angled

Curvatures of this depth will not create focus problems with most multiple-image presentations – the depth of field of projection lens will be more than enough to permit sharp images. (Depth of focus is another matter of course; a slight movement of the projection lens in relation to the transparency will have much more effect on screen image sharpness than will the curvature of the screen.)

When matte screen surfaces are used with curved or angled screens, there may be a problem with stray light reflected from one end of the screen to the other. That is, a bright image projected on one end of a curved screen or on one angled screen may reflect enough light to degrade the image on the opposite end of a curved screen or on the other angled screen – particularly if there is a dark image there. The problem does not usually occur with non-matte screen surfaces; it can be reduced with matte screens by embossing the surface to make it slightly rough or by making the surface of a moderately rough material, such as plaster.

If a deeply curved screen is used or if flat screens are very sharply angled to each other, there may be increasing problems with light from one projected image degrading the other image. Using deeply curved screens may also cause focus problems, and noticeably different magnification as between image ends and the centre of the image. Keeping the screen radius close to the 1 to 1.5 times viewing distance mentioned above will normally prevent these problems.

Placement of the projectors will be determined by the angle of flat screens to each other or the radius of a curved screen. Set up your projectors so that the projection beams are at right-angles to the screens or, if a curved screen is used, to the chords of the image areas. If the crossover point of the projection beams is within the room with the projectors at the back of the room, the left projector should project the right image and vice versa (see Figure A.6). On the other hand, if the crossover point occurs near the back of the room or outside the room, it is usually preferable to have the right projector on the right screen and the left projector on the left screen as shown in Figure A.7.

Most of the time multiple images should be projected to be about the same size and brightness. For instance, the images may be all slides, preferably well matched in size, brightness and contrast. If dissimilar images are combined (for example, an overhead projector image with a projected slide or a motion picture with slide), try to get them close to each other in both size and brightness – unless they are especially designed for some effect requiring different sizes.

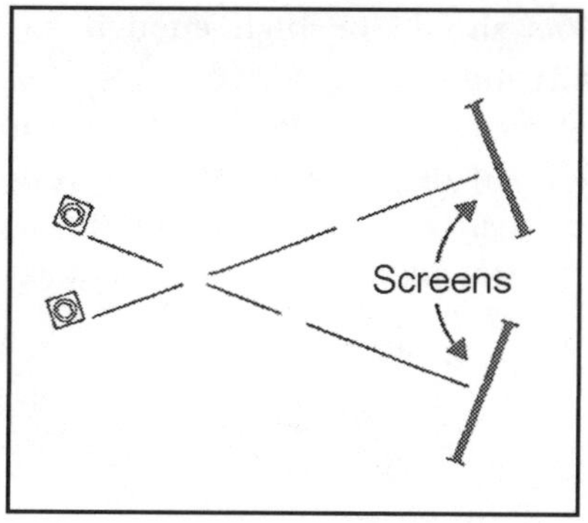

Figure A.6 Placing of projectors – crossover point within the room

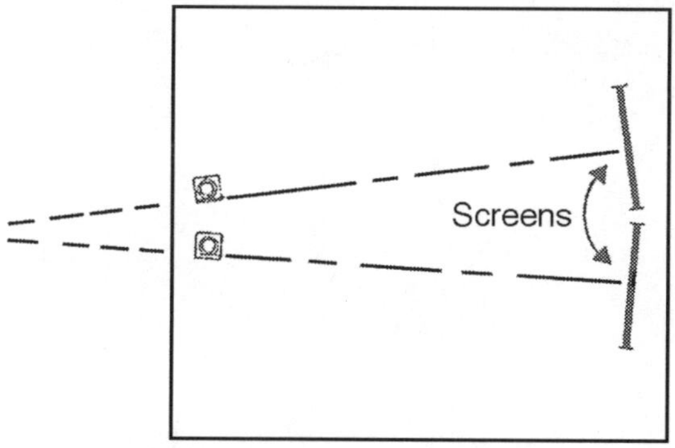

Figure A.7 Placing of projectors – crossover point outside the room

Loudspeaker location

Sound quality is usually quite satisfactory if the loudspeaker is near the screen and high enough to be seen by everyone. Low placement causes loss of intelligibility for all except those in the front row.

Where acoustics are poor, a corner location or extra speakers placed towards the back of the room may help. Extra speakers should ordinarily be aimed towards the back of the room to avoid interference between separate units.

Proper tone-control adjustment can greatly help in any room. In an acoustically poor room, the best intelligibility is usually achieved with the tone control near maximum treble. These settings do not give the most pleasing tone for the reproduction of music, but they do reduce reverberation and make speech more understandable.

Projector location

For an undistorted image, the projector lens should be a line extended at right angles, vertically and horizontally, from the centre of the screen surface. For overhead and opaque projectors at the front of the room, this usually requires that the top of the screen be tilted toward the viewers. Projectors should be high enough so that their beam will clear obstructions, such as the heads and hats of viewers.

Excessive upward or downward projection angles will cause keystoning of the image unless the screen is tilted. A slight degree of keystoning is not objectionable. With most lenticular screens, a change in screen angle that introduces only slight keystoning can often improve the image by rejecting stray light and increasing image brightness as viewed by the audience.

PART II

Handling the Media

Introduction to Part II

People, even prominent people, whose views are sought out by journalists are all too often unprepared and unskilled in dealing with media personnel and lack the confidence to look for and exploit opportunities that are available to them. This is at a time when the public is accustomed to a high standard of presentation by media professionals. These journalists are not only skilled at their job, but they have all the appropriate equipment and are backed up by support staff ready to prompt them through their concealed earpieces if they suffer a momentary lapse.

The probability of being interviewed has increased substantially; journalists are constantly seeking out those who should have answers to key issues and problems, so that what they say can be conveyed to their viewers, listeners or readers. People who are likely to be invited or confronted by the media need to be equipped with the techniques and skills to handle interviews, to make presentations to the media and to deal with questions raised by journalists. For the media, bad news is always good news, so it is essential for those who have to deal with or announce adverse news, to be able to do so first and fast, rather than be taken by surprise.

Those in responsible positions are nowadays expected to be able to conduct press interviews and to be able to broadcast: but if such media opportunities are to produce benefit and not disaster, you must know how to speak well and present your case, and how to deal with the questions and comments put to you, fluently, confidently and persuasively. Above all, what the television viewer and radio listener sees and hears must be believable.

By learning a few straightforward rules, by developing skill and technique, you can become a creditable spokesperson for your organization. Fluency and competence in handling television, radio and press interviews is an essential management skill. Such skills must be developed throughout a company's senior management. Do not make the mistake of a newly appointed chief executive of one of Europe's top ten pharmaceutical companies where there was a programme for training the company's top one hundred managers in every subsidiary company across the world in how to handle the media. He scrapped it – and decreed that only he would deal with any questions from the media. Bad press comment and disaster followed. Eventually the company was taken over.

Part II of this book is designed to provide a foundation of knowledge and techniques for handling the media. With its help you will be able to

- set clear objectives for media interviews and, at the same time, understand those of the media
- control media meetings and interviews so that televised, broadcast or published features do not contain something you said in an unguarded moment and now wish you had not

- handle sensitive issues so that the correct facts and impressions are conveyed to the public
- prepare, structure and conduct all types of media interview.

On pages 86–87 is a questionnaire covering your perceptions of the media (Figure I.1). It has been designed so that you can record not only your perceptions, but also your comments on

MY EXPERIENCE AND PROBLEMS FACED IN HANDLING THE MEDIA

PRESS	RADIO	TELEVISION

My Percepions of the Media

For each medium tick the rating that most closely represents your own perception.

PRESS

	VERY	SOMEWHAT	INDIFFERENT	SOMEWHAT	VERY	
1 is sympathetic to my organization/industry	○	○	○	○	○	is antagonistic to my organization/industry
2 is objective	○	○	○	○	○	is biased
3 gives fair treatment	○	○	○	○	○	is manipulative
4 is well prepared	○	○	○	○	○	is superficial
5 shows undersanding of my organization/industry	○	○	○	○	○	lacks understanding of my organization/ industry

Figure I.1 Media perception questionnaire

RADIO

	VERY	SOMEWHAT	INDIFFERENT	SOMEWHAT	VERY	
1 is sympathetic to my organization/industry	○	○	○	○	○	is antagonistic to my organization/industry
2 is objective	○	○	○	○	○	is biased
3 gives fair treatment	○	○	○	○	○	is manipulative
4 is well prepared	○	○	○	○	○	is superficial
5 shows undersanding of my organization/industry	○	○	○	○	○	lacks understanding of my organization/ industry

TELEVISION

	VERY	SOMEWHAT	INDIFFERENT	SOMEWHAT	VERY	
1 is sympathetic to my organization/industry	○	○	○	○	○	is antagonistic to my organization/industry
2 is objective	○	○	○	○	○	is biased
3 gives fair treatment	○	○	○	○	○	is manipulative
4 is well prepared	○	○	○	○	○	is superficial
5 shows undersanding of my organization/industry	○	○	○	○	○	lacks understanding of my organization/ industry

Figure I.1 (continued)

any experiences you have had of being interviewed by the media and any problems you may have encountered or perceived. When you have completed Part II you will be asked to fill in the questionnaire a second time. You can then compare the two versions to see if your perceptions have changed as a result of a better understanding of what journalists do and how you can handle them.

I suggest that at this point you make two photocopies of the blank questionnaire. Complete one of them now and keep the other somewhere safe until you reach the end of the book.

When you have studied Part II you will have a better appreciation of the world of the media. Treat journalists as we all, as customers, wish to be treated. Try to understand their needs and objectives and you will succeed most of the time in communicating your point not only to them but through them to the publics you seek to influence – consultants and professionals working in hospitals, GPs, those in Primary Care Trusts and of course those who ultimately use your products, patients.

CHAPTER

8 Facing the Media

The decision whether or not to talk to the press, appear on television or take part in a radio programme is like any other management decision and should be made in the same way. When you are asked to give a press, radio or television interview, always weigh up the pros and cons of doing so.

Factors in favour

Here are a number of obvious reasons for agreeing to be interviewed. You will no doubt be able to add others relevant to your own or your company's situation:

- provides a public relations opportunity
- enhances your organization's status in the community
- provides free promotion for your organization
- promotes a positive image of your organization to potential staff
- provides an opportunity for personal interview practice
- provides an opportunity to inform or influence public and professional opinion
- provides an opportunity to correct or prevent misunderstanding about the pharmaceutical industry
- strengthens staff motivation
- provides an opportunity for self-projection to increase the likelihood of being asked again
- provides an opportunity to put your organization's case versus that of competition
- provides an opportunity to change public attitudes
- provides an opportunity to influence specific target groups such as financial institutions, scientists, young people, government departments.

Factors against

Now consider all the reasons for not doing a press, radio or television interview:

- it could prove an unpleasant experience
- time pressures
- lack of time to prepare; never under-estimate how long this takes
- you are not the right person to do it
- danger of performing poorly

- you do not want to disclose your organization's policies or plans at present
- you do not want to be exposed to being browbeaten or made to look foolish
- organizational culture
- pressure from your colleagues
- danger of your views being distorted by the media
- loss of the comfort of anonymity
- possible threat to your career
- bad timing for stock market.

Co-operating with the media

There is much to gain and nothing to lose by co-operating with the media and building up a fund of goodwill with journalists. Often, journalists will get in touch with you. They go on a 'fishing' expedition, searching for information or the names of reliable people who could take part in a programme or provide specialist advice. All journalists keep a 'little black book' of best sources of information. Becoming a journalist's source of accurate information or being able to recommend useful people for them to contact, is a valuable asset.

When journalists have tested your recommendations and found them reliable, your advice will be sought again. When you recommend other people to take part in a particular programme, if you feel that you are not the best person, do not worry that you may be missing an opportunity. Better to forgo the opportunity this time than to be the wrong person and risk looking foolish. An occasion will arise where, when asked who is the best person to do such and such an interview, you can say, 'Well, you are talking to them – I am'.

Never underestimate the benefit that can result from taking part in a programme. An example of this occurred when I was directing a training programme for a group of senior medical sales representatives employed all over the world by the pharmaceutical division of the former Imperial Chemical Industries (now AstraZeneca), at their management training centre in Kingston-on-Thames. It so happened that Terry Wogan had persuaded the then chairman of ICI, Sir John Harvey-Jones, to be interviewed by him on television one evening that week.

No senior member of this company, let alone the chairman, had ever appeared on a television programme before. Sir John is very fluent, very amusing and known for his colourful language and dress. He was also, as he often demonstrated, a superb ambassador for his cause and his company. The effect of his television appearance on the group I was with was electric. In the space of ten minutes, Sir John did more for the morale of the entire ICI staff than all the motivational talks that top management could have given. It also happened at a time when the very survival of the ICI group was at stake. One division alone had reported a loss of over £260 million.

When a news story breaks, a reporter wants to speak to a representative of the business concerned, and not be fobbed off with someone intent on putting up a smokescreen or feeding them a bland public relations line. If the media suspect a cover-up, they will camp on your front lawn until they get the facts from the 'horse's mouth'.

If you decide against co-operating with the media or decide not to appear, then you would be well advised to give the journalist, radio or television company a carefully prepared statement on your point of view, if you can offer one that will advance your case. Then, if reference is made to your non-appearance, you have at least provided material that can be quoted. Moreover, if it is not used or is misquoted, you have a legitimate basis for complaint, using such methods as letters to the press or television 'answer back' programmes, e-mailed views and similar radio versions. You always have to balance the decision not to co-operate against the danger of providing the media, or your competitors, with an opportunity to attack you. Sadly, however good your reasons, when the news reporter makes the dreaded comment, 'We invited a spokesperson from . . . to appear on this programme but the company declined', countless thousands of viewers and listeners can almost be heard saying, 'What are they trying to hide?'

If you do decide to co-operate with the media, or when there is no option but to do so, then your preparation must be thorough: *prepare, prepare,* and yet again, *prepare yourself.* Never rely on your ability to improvise. Remember also the great difference between written and spoken words. What may read well on the printed page may not sound so convincing when spoken.

CHAPTER 9 Dealing with the Press

When you deal with the press, either about an article or in response to a request for an opinion, it is important to understand clearly the uses to which any information you provide may be put. The aims and interests of journalists and those of your company or organization do not usually coincide. A journalist's first loyalty is always to his or her readers, not to his or her sources of information. Conversely, your first loyalty is to your company or organization and its reputation.

Beware of making assumptions about other people's motives. You may be motivated by money. A great many journalists are not. Their ambitions are often quaint and complex. The search for the truth and exposing it can be a compelling force for many.

Journalists are a part of the hard-nosed business of publishing. You may see them as a young, modestly paid, man or woman, but remember that the journalist is always in control. It is what they report back to their publications that gets into print.

Always clarify with journalists how the information you provide will be used. You must distinguish between *supplying* information and *using* information as a means of winning the journalist's loyalty to your cause. If journalists forget to whom they owe their first loyalty, the price could be their job.

Beware of the freelance journalist who rings you up and asks for an interview. Often they have no status and are hoping that you can give them some sensational or scandalous information around which they will be able to write a saleable article. Always ask such journalists to show you written confirmation of the article they have been commissioned to write, the name of the publication in which it is to appear and the date when it will be published.

It is better to supply information that clarifies a situation to journalists than to refuse to do so. This prevents them from writing that when the company in question was invited to make a statement, it refused to comment. This type of remark can only create a suspicion that you have something to hide in the minds of readers.

The conventions

When you talk to the press, a number of well established conventions govern how what you say may be reported. Always stipulate, before any interview takes place, the basis on which you are prepared to be interviewed.

ON THE RECORD

Anything you say 'on the record' may be attributed to you as a direct quotation. Unless you deal with this matter clearly and firmly before the interview starts, then the journalist will *assume that everything is on the record.* After such an interview, the journalist is at liberty to use any information you gave them without clearing it with you first. The following points apply to 'on the record' interviews:

- You have no automatic right to a preview of any written material in which you are quoted, although you should always ask anyway. In the case of sensitive information such as complex drug formulae, financial details or dense statistical data, you should insist on being able to check them, because the consequences of any misprint could be serious. This approach will usually secure the journalist's co-operation.
- If you prefer not to be named, tell the journalist that although what you say is on the record, you wish it to be attributed to a spokesperson, not to you by name.
- Sometimes you may want to give a journalist some background information, the better to understand what you are about to say. In that case state that you wish to offer an *unattributable view* which is not to be quoted, nor attributed to you.
- If you do not wish to be quoted at all, but are willing for the information you provide to be incorporated in an article, tell the journalist that you are *speaking unattributably*.

OFF THE RECORD

Never speak to a journalist 'off the record'. If you want to be sure that something remains off the record, do not tell it to a journalist. Keep away from them. You cannot attempt to collude with someone whose loyalties, aims and motivations are different from yours. This does not mean that journalists are not to be trusted under any circumstances. You just have to remember how they earn their living and that information provides the foundation on which they build their articles.

Avoiding misrepresentation

Although there is always a risk of being misrepresented by the press, this danger can be minimized by bearing the following points in mind.

BE ACCURATE

Make sure that what you say is accurate and as factual as possible. Facts can be verified, opinions cannot, so stick to the facts, do not proffer opinions. If you are asked a factual question which you cannot answer, do not hesitate to say so. If the question is an important one for your company or organization, then add that you will do your best to find out. The temptation to give an opinion in such circumstances should be resisted. It will get you into trouble.

USING A TAPE RECORDER

Some journalists prefer to conduct an on the record interview using a tape recorder rather than taking down your comments in shorthand. Sadly, few journalists write shorthand nowadays. I admit to using a tape recorder for my interviews with people I am going to write about. It gives me the freedom to look at the person I am interviewing and makes for a much more relaxed atmosphere.

Clarify the basis on which the tape will be used, although as a rule there is no reason why you should object. When a journalist uses a tape recorder for an interview, some interviewees feel that they should produce their own tape recorder and make a separate recording of the conversation. Although a journalist probably would not object, it does not help the two-way flow of the interview or generate an easy environment in which to exchange information.

COMPLEX INFORMATION

If your interview with a journalist is likely to involve figures or complex data, it is best to send such information in advance of the interview so that it can be studied beforehand rather than risk delay at the meeting. Be sure to spell out unusual names. And watch out for mistakes in noting down even common words. There is a world of difference between: 'Our objectives for this clinical trial ...' and, 'Our objections to this competitor's clinical trails are ...'

DON'T FEAR SILENCE

During a interview there may be a number of pauses whilst the journalist is digesting what you have just said and framing the next question. Do not let the silence bother you.

Silence and how it is handled is the interviewer's problem, not yours. Make the statement you want to make or answer the question as you planned and then *stop talking and wait* for the next question. You may make the mistake of assuming from the silence that the journalist is expecting you to say more. If you have no more to say but feel you should speak, what you then say may be unwise or even counter-productive in terms of your objectives for the interview.

STOP TALKING

If, following an interview on the record, you complain that you have been misquoted, the journalist may play back the tape recording or read out a transcript and prove that you did say what you now dispute. At this point, it is no good saying, 'Well that's all very well, but what I meant to say was...' You will be reminded that you did agree to speak on the record and therefore anything you said was perfectly legitimate for the journalist to use.

When speaking to a journalist on the record, beware when they close their notebook as they prepare to leave. You inwardly give a sigh of relief and may let yourself relax. This is the moment when interviewees make the most damaging and quotable comments. You may think it unfair but you are 'on the record' until the journalist has finally departed.

DON'T WAGE WAR ON THE PRESS

If you become incensed by something that has been printed about you or your company, remember that declaring war on the press is always futile, be you ever so powerful or rich. Never pick a fight with someone who buys his ink by the barrel. Even the late Jimmy Goldsmith, one of the world's richest people, could not stop the 'Golden Balls' articles about him appearing in *Private Eye*.

PRESS DEADLINES

If a journalist is working to a deadline, find out what it is. If it is too close for you to be able to obtain the necessary answers to questions raised, say, 'I'm sorry. I won't be able to get answers to you before that time. Would you still like me to find out?'

DON'T SPECULATE

Never speculate. If a journalist tries to tempt you into speculation with prompts like, 'Do you think that such and such is probably the case?' decline to answer or even to indicate whether you agree.

Don't be fooled by journalists who try to insist on the importance of their job as champions of the public conscience or who argue that people have a right to know what is going on. Who says so? The press!

ENSURING UNDERSTANDING

Few journalists know what they are talking about or what they are planning to write about. Their skill lies being able to collect facts and then communicate them in cogent and attractive language that people will want to read. Journalists need educating. So when imparting information, ensure that they have fully understood. Test their comprehension by asking them to feed back what you have said.

TEST UNDERSTANDING

If you do not feel confident that you understand something that a journalist wants to discuss with you, be careful. Preferably do not discuss it.

ADMIT IGNORANCE

If you answer a journalist's question with a statement which you do not fully understand they will probably realize it and this could undermine the rest of the interview. You will have lost credibility.

If you face this problem, admit to the journalist that he or she is asking about an area you do not understand. Offer to find someone with the specialist knowledge required who can call the journalist and explain it competently. To ensure that you are clear about the brief to provide a colleague, ask the journalist to let you have the questions so that you can write them down. Before you pass them on repeat the questions to the journalist to ensure that you have written them down accurately.

Features for publication

The preceding sections of this chapter relate mainly to stories appearing in print and describe how you should *react* to approaches made to you by the press. But there are also important opportunities to be *proactive* which you and your company should be ready to exploit when the time and the material are right.

You can obtain positive publicity and goodwill for your organization and its activities from feature articles. Do not look on this as a glorious opening for you to achieve that lifetime ambition to become a writer. An outside expert is rarely asked to write an article for a newspaper or magazine. However, if it does happen, do not be fooled into thinking that 2,500 words can be rattled off in a couple of hours. The more sensible option is using a professional journalist.

Usually an article will be written by a journalist, based on an interview and other relevant material supplied by one or more outside experts. A journalist writing a comparative piece will want to talk to sources other than you and probably, if you work in a commercial company, to some of your competitors.

From whatever material you supply, either at an interview or in written form, the journalist will select what he or she thinks will best meet their readers' needs and expectations. The final text will also reflect not only the journalist's own style but the way in which the commissioning publication likes its material to be expressed. Although the result may not always be what you would have written yourself, remember that the journalist is a professional communicator and you should respect their expertise.

If you give an interview, as I have said many times, prepare the material and yourself thoroughly. Before granting a journalist an interview, it is worthwhile asking yourself whether it will be to your benefit to do so. Do not be flattered by the invitation, especially if it is your first. Ask yourself the following questions:

1 DOES THE ARTICLE THIS JOURNALIST IS PROPOSING TO WRITE SOUND INTERESTING?

If the topic is say, 'Opportunities for computer programmers in pharmaceuticals', *BEWARE.* The idea may have been thought up to fit the space between the advertisements in the computer vacancies section of the magazine or newspaper. Always be on your guard against public relations puffs used to separate advertisements. There are magazines whose whole annual programme is centred on such topics as Conference and Exhibitions; Sales Incentives and Promotions; and Marketing Research.

When you analyze the articles that appear each month, they contain one or more relating to the specialist activity for which the publisher hopes to sell advertising space. The *Financial Times* weekly and monthly supplements have been built round this strategy. Unless you want to attract computer programmers or whatever specialist jobs are being featured, it is not worth co-operating with the publication.

Advertorials

An advertorial is advertising presented in the guise of editorial material. Some UK national newspapers feature advertorials, for example, the *Financial Times* and *The Times*, and a number of monthly magazines publish supplements covering service industries like management consultancy and medical insurance. The pharmaceutical industry specialist publications publish advertorials regularly.

I am often asked to write advertorials for companies in the pharmaceutical industry and those who service it. But I accept few such invitations because those issuing them are not prepared to meet the stringent rules I lay down for doing them. Too many of those I read are little more than unctuous effusions with the advertiser's name repeated in every sentence, which defeats the objective for which it was written. They repel the reader after one paragraph. Advertorials form part of a company's promotional expenditure and can be expensive: a five-page advertorial with artwork can cost between £12,000 and £30,000.

An advertorial should not be commissioned without a great deal of thought. The following factors should be borne in mind:

- An advertorial should have a clear and measurable objective such as to promote a product or service or to position a company's reputation and performance in its field of operations to interest another seeking to merge with it or take it over. This will enable the results of the advertorial to be measured.
- It should show how it meets a need or solves a problem; the benefits a potential buyer would get; and the action to be taken by the reader.
- To be effective, an advertorial should feature the company's weaknesses as well as its strengths. No company is perfect and an advertorial that claims perfection will merely antagonize its readers.
- An advertorial should, ideally, be included in the main section of a newspaper or magazine, rather than form part of a separate supplement. If the advertorial looks too much like an advertising or public relations puff, it will be discarded unread. I always follow up those I write, to find out what results were achieved by those who commissioned them. Those in separately published supplements rarely produce any measurable response, unlike advertorials appearing in the main sections of a publication.
- Because of the cost of advertorials, the results should always be measured. One effective method is to ask the publisher to produce reprints and mail them out to targeted readers with specific needs in the area featured in the advertorial. Then measure the outcomes from following up the recipients.

2 DOES THE JOURNALIST APPEAR TO KNOW ANYTHING ABOUT THE SUBJECT?

Despite what I have said previously, some journalists become deeply knowledgeable about a particular industry or activity. One or two pointed questions about how he or she plans to tackle the subject will soon indicate the depth of their knowledge or ignorance.

Do remember that knowing nothing about a subject does not necessarily disqualify a journalist from writing a good article about it. However, it does mean that those who are interviewed need to take great care over how they impart information and to be aware of the dangers of using specialist jargon. The journalist might not understand but refrain from admitting ignorance.

I recall an occasion when I was one of a number of experts interviewed by a freelance journalist about the pharmaceutical industry. She decided to send me a draft of the whole article she proposed to send in to the publication in which it would be printed. Some of the people she interviewed had given her information which anyone in the pharmaceutical industry would realize at once was wrong. Fortunately, she accepted my advice on matters of fact, re-checked with her sources and made corrections to the copy before it went to press. I see examples of this type of ignorance by freelance journalists daily in print.

Use of tape recorders

For an article interview, if the journalist wishes to use a tape recorder, allow it unless there is a particular reason for objecting. Tape recorders are an essential part of any journalist's equipment. If you are concerned about the danger of being misrepresented, ask someone else in your company to sit in on the interview.

Telephone interviews

For feature articles, face-to-face interviews are better than telephone conversations. But if the deadline for the article is close, you may have to no alternative but to give an interview by telephone.

Be careful about talking off the cuff about sensitive or political issues. Where time allows, call the journalists back after you have had time to think about the subject and what your responses would be to potential questions. As a general rule, it is wise to grant a telephone interview for an article only if the subject is straightforward and the interview itself is likely to be short.

Face-to-face interviews

If the subject matter is complex and the journalist is to interview you in person, see whether you can arrange to have some published material available. This can be taken away and studied and may also serve to draw attention to the correct spellings of unusual or highly scientific words.

Here your preparation for the interview should include thinking about what documentation might help the journalist. Consider your company's annual report, reprints of articles or medical or scientific papers presented to a conference, the organization's public relations files and cuttings books. All these should be to hand, with photocopies ready if you plan to give the journalist material to take away.

Previewing the text

Journalists work to different deadlines, according to the frequency of the publication for which they are writing. There is nearly always more time available for the preparation of feature material than there is for news copy for publications like a national daily that has to be ready for printing ('put to bed') by, say, 7.00 pm on the day before publication. Consequently, you should always confirm before agreeing to an interview that you will either be shown or read over what the journalist writes about you or your company before it is published.

It is best if you can actually see the text. If time is too short for this, the journalist should agree to call you when the article is complete and read over what they have written. E-mail and fax machines now enable the actual draft to be seen as soon as it is written.

If you have an opportunity to comment, restrict yourself to matters of fact relating to your own interview and what you said. It is unlikely that a journalist will fax, e-mail or read to you the other sections of their article. But if this happens, you have no right to criticize them, still less to suggest changes. Similarly, if the journalist has expressed an opinion with which you disagree, that is their prerogative. On the other hand, if you feel that the opinion expressed is based on a misunderstanding of the facts as you gave them, then point this out firmly but quietly. Showing that you are annoyed or angry could provide the journalist with ammunition to give an unfavourable complexion to that part of the article that bears upon your interview. Only insist on a change if there is a factual error, or if you have been quoted as saying something you did not say.

Remember that you have no right to withdraw something which you did say but, on reflection, wish you had not said. Perhaps you blurted it out during one of those journalistic silences!

If you are sent an article in draft form and it includes comments from a variety of sources, some of them perhaps made by your competitors about you, you have no right to request a change to that text unless what is said is factually wrong and you can prove it to be so. You may not like the views expressed in the article by other people, but that is the price we pay for a free press.

You should, however, not hesitate to point out any errors of fact in parts of the article not concerned with what you said if you believe that their inclusion could invalidate the whole thrust of the article. If this occurs, suggest quietly to the journalist that they check the facts with the source(s) from which they were obtained.

Sometimes, despite all the checks and references back to you about copy, the article that appears, or that part of it that refers to your interview, is wrong. Do not jump to the conclusion that it was the journalist's fault. The story may have been changed, cut or reworded by a sub-editor.

Finally, if you are unhappy with what appears in print, try not to take it too seriously. Newspapers, especially, are largely ephemeral, their contents soon forgotten by most people even before the next issue comes out. Only if damaging untruths or misinterpretations are printed should you seek redress. It is always salutary to remember that today's newspaper lights tomorrow's bonfire.

Some real-life case studies

The five case studies that follow provide an ominous warning that even prominent people, with all the professional advice at their disposal, can make basic mistakes in their dealings with the press.

CASE STUDY 1 THE WELLCOME FOUNDATION PRESS CONFERENCE

The Wellcome Foundation, owned by what is now GlaxoSmithKline, was before it was taken over a private company wholly owned by The Wellcome Trust, a charitable foundation set up to provide funds for medical research.

In the 1980s The Wellcome Trust decided to float 25 per cent of the company on the Stock Exchange. During that decade it had brought to market Retrovir and Zovirax, two drugs which in combination were the only treatment available to patients suffering from HIV infection. These drugs did not cure AIDs; nothing could. But they helped to prolong the lives of those with AIDS. Wellcome was thus very much in the public eye, and consequently a constant source of news stories for journalists and health correspondents of the national daily newspapers, not to mention BBC and ITV television.

For both reasons, the Wellcome board decided that it would be sensible to identify the top management throughout its worldwide operations and put them through a workshop to develop their skills in handling the media. I was asked to design a workshop to achieve this management training objective. My research included attending the annual press conferences, which preceded the annual general meetings, to obtain firsthand knowledge of how they were conducted. These press conferences, chaired by the Wellcome Foundation's chairman, were sometimes quite dramatic. Not least because the then chairman thought that all his senior colleagues should be trained to handle the media, but not him. The following exchange, part of a question-and-answer session at one of these annual press conferences, illustrates how wrong he was.

Guardian health correspondent: 'Could you please confirm that the annual cost of treating a patient with AIDS is £6,000?'
Wellcome Foundation chairman: 'I am sorry, I cannot answer that question.'
Health correspondent: 'Why not?'
Chairman: 'Because it is not one of the questions I have down to answer.'
Health correspondent: 'I will repeat my question. What is the total annual cost of treating an AIDS patient? Is it £6,000?'
Chairman: 'Well, if you insist, I will ask the finance director to deal with that question'.
Finance director: 'The annual cost is not £6,000, it is £4,500'.
Health correspondent: 'And how much are the distribution costs?'
Finance director: 'The annual distribution cost is £1,500'.
Health correspondent: 'If I add £4,500 and £1,500 the total is £6,000?'
Finance director: 'Well, yes, that would be correct'.
Health correspondent: 'So my original figure of £6,000 per annum was in fact correct?'

Neither the chairman nor the finance director replied to the last question from the *Guardian* correspondent!

CASE STUDY 2 THE TRANSLATION TRAP

Our second case study also involves The Wellcome Foundation. The president of the group's Japanese subsidiary company was a Scotsman who spoke fluent Japanese. I gave him one day's private coaching in the techniques of handling the media. When we came to dealing with the press in Japan, he told me that because he spoke fluent Japanese he felt able to handle questions from the press in their own language. What did I think about that? Under questioning, he agreed that he often found himself at a loss how to answer, and would like to have more time to think before replying.

I told him to stop this practice. He should agree to take questions, but ask that each question be translated into English so that he could the better understand them. The reason I suggested this was so that he would have more time in which he could listen to each question in the Japanese language which he understood and so be forewarned of its contents. Because each translation took a long time, he could weigh up the implications of what he was being asked and work out how to answer. Later he told me that he was able to deal with the questions much more effectively.

Former Prime Minister Harold Macmillan bungled badly when he assumed that, by talking to General de Gaulle in French, he would be able to win his support for Britain's application to join the European Common Market. In this he failed completely.

CASE STUDY 3 PAYING THE PRICE FOR IMPULSIVENESS

Gerald Ratner was once chairman of a 238-strong jewellery chain that bore his name. On 23 April 1991 he had been invited to make one of the keynote speeches at the annual convention of The Institute of Directors (IOD), held each year at the Royal Albert Hall in London. I had been invited to attend it as a member of the press and was sitting with other journalists and BBC camera crews just below the speaker's platform.

Gerald Ratner was due to speak in the afternoon following the traditional luncheon box feast provided for all delegates. This always includes a generous supply of wine. So he had to contend with one of a speaker's most formidable challenges – how to keep delegates awake and listening after lunch. Some snore. Others have, over the years, developed the ability to sleep with their eyes open. Conscious of this problem, Ratner once or twice departed from his prepared script to make some impromptu remarks which he hoped would hold the attention of the audience. He referred to some of the products in his jewellery outlets in the now infamous phrase 'total crap'. This was not in the press handouts of his speech, advance copies of which had been given to journalists including myself.

In the days that followed, sales in his shops plummeted and he was forced to launch an advertising campaign, estimated to have cost the group between £250,000 and £500,000, in an attempt to offset the effects of his widely publicized comments implying that some of his merchandise was cheap and tacky. Outside the Christmas season, this retail group had never previously advertised in the national press. On 23 May 1991 the advertising and promotion magazine *Campaign* reported that 'sales slumped despite a desperate celebrity-endorsing ad campaign rushed out the following week [after the IOD convention]'.

The after-shocks were felt by Ratner, his company and the shareholders. On 1 April 1991 Ratner ordinary shares stood at £1.80 each, valuing the company at £522 million. On 7 June 1991 they stood at 33.5 pence and the company was worth £98 million. Uttering those two words reduced the value of Ratner's shares by £1.465 and the value of the company by £424 million.

Ratner himself was forced to resign from the chairmanship of his company and to leave it altogether. Then, to have any hope of regaining a respectable status in the retail jewellery market, the company had to change its name. The company that was once Ratners is now called Signet and on 9 August 2002 Signet's ordinary shares stood at 86 pence.

This is a dramatic example of self-induced damage brought about by lack of thought for the dangers involved in making remarks 'off the cuff'. They may be intended to galvanize an audience which the speaker thinks needs waking up. But the press considered them ill-judged and flippant. And the stories they filed for their newspapers and for their television and radio stations showed this only too plainly.

Our last two case studies concern two directors in the pharmaceutical/bio-science industries whom I interviewed for articles in *The Director* and *Pharma Times*. Neither interview was confrontational, but both demonstrate the value of open-ended questions in obtaining descriptive, informative feedback. Open-ended questions begin with one of the following words: Who?, Why?, What?, Where?, How?, When? Unlike 'closed' questions – 'Did you call on Dr Green?' or 'I imagine you were against that policy?' – to which the answer can only be 'yes' or 'no', open-ended questions can never be so answered.

The skilled interviewer sees to it that the interviewee talks for ninety per cent of the time, while he or she talks for only ten per cent.

CASE STUDY 4 NO PLACE FOR ENTREPRENEURS?

Sir Christopher Evans, chairman of Merlin Biosciences, gave a keynote speech at the 2001 annual convention of the Institute of Directors. Extracts from my report of his speech are reproduced below, together with my subsequent press interview with Sir Christopher. The whole was published in *The Director* magazine under the title: 'Pharma Industry – No Place for Entrepreneurs'.

At the last two annual conventions of the Institute of Directors, Jan Leschley, followed a year later by Sir Christopher Evans, said the pharmaceutical companies are unattractive places for entrepreneurs. John Lidstone summarizes their views.

Entrepreneurs + money = faster cures! Perhaps. But not in pharmaceutical companies. When you hear claims that it takes £350–£500 million to research and then bring a new chemical entity to market, you could be forgiven for questioning my opening statement. Except that it was not made by me, but by two doyens of the industry who spoke successively at the 2000 and 2001 annual conventions of the Institute of Directors, Jan Leschley making his valedictory speech as chairman of SmithKlineBeecham, and Sir Christopher Evans, chairman of Merlin Biosciences.

First, Jan Leschley threw down a challenge for the industry just as he was leaving it following the merger between SmithKlineBeecham and Glaxo Wellcome. He told his audience that he had given his twenty-eight-year-old son $2 million to start a new venture in America. Six months later, having built it into a credible commercial entity, he sold it for $200 million. 'How can we attract talented twenty-eight-year-olds into the industry to do what they want to do?' he asked.

Sign-posting the way ahead, Sir Chris Evans said, 'Britain is number two in the world in bioscience. We produce all sorts of great inventions, discoveries and innovations. We need more and more of that fabulous science. We need to bring more of our brilliant scientists through our schools, our universities and into our research laboratories. We need to invest large sums of money. And also we need the entrepreneurs. We need superb leaders, flexible, entrepreneurial business managers, and inspirational, motivational leaders.'

Searching questions

JL: 'Last year, Jan Leschley spoke about the problem of attracting entrepreneurs into the pharmaceutical industry, not least because there is no payback for them for years. Yet this industry must innovate or die. How are you, in bioscience, going to attract the entrepreneurs you so desperately need?'

CE: 'Funnily enough, some of them are latent, frustrated entrepreneurs from the pharmaceutical industry. They come to us and have a passion almost to prove a point. But I have to admit, it is difficult to attract them.'

JL: 'How do you cope with managing entrepreneurs? After all, they are usually brilliant in thinking up new ideas, but hopeless at managing the results of them.'

CE: 'You can build disciplines around scientists once you bring in venture capitalists. They tend to be very experienced in providing bedrock management to underpin these geniuses. You have got to get the right balance between creative scientists and management.'

JL: 'Jan Leschley justified the merger between Glaxo Wellcome and SmithKlineBeecham on the grounds of size plus savings of about £350 million, barely sufficient to fund one new marketable chemical entity. How are you going to raise the all-important money you need?'

CE: 'It is true that at present in the UK, we have not got enough cash going into enough companies. The problem is that the people with the cash to invest do not trust the management of biochemical companies to perform. In the US, the attitude is, "Let's go for it". In Britain it is, "Let us see your product, then we will think about investing."'

JL: 'How likely are bioscience companies to go the same way as the pharma companies, who are becoming larger in size, but fewer in number?'
CE: 'Yes, I think that you are seeing this already with companies like Shire, Celltech, and Powderject. We've got a myriad of companies in the £50 million-plus size, but they are just not big enough. We need more companies with critical mass. Having said that, there is a safety in numbers. We are seeing a cycle in the industry consolidation, leaving space in between for new companies.'

Reprinted extracts from *The Director*. Copyright John Lidstone 2001.

Following his election to the office of President, The Association of the British Pharmaceutical Industry (ABPI), in April 2002 Dr John Patterson, director, AstraZeneca, agreed to be interviewed by me for *Pharma Times*. The following extracts from this lengthy interview show once again how open-ended questioning of the interviewee works. Dr Patterson made a number of frank comments in an interview which was not intended to put pressure on him, but simply to encourage him to talk openly about critical issues facing this important industry.

CASE STUDY 5 THE PRESIDENT SPEAKS

'Two years is a short time as president of the ABPI. What have you chosen as the one, or at most two, objectives to achieve?'

'For many of my predecessors, the presidency has been their final, virtually full-time job before slipping into glorious retirement. Consequently they have had the time to pursue their chosen theme. I am going to be more of a non-executive chairman than a day-to-day hands-on chief executive. Three areas will be important.

The Pharmaceutical Price Regulation Scheme (PPRS), pricing and associated issues will continue to be a high priority. Second, access to medicines. NICE (the National Institute for Clinical Excellence), the proposed increased switching from POM (prescription-only medicines) to P (pharmacy) and associated matters. Thirdly, relationships. We need to improve our relationships with our customers, whether they are doctors, pharmacists or even patients.'

'How independent, politically, is NICE, given that it was created by this Government?'

'Currently decisions are taken behind closed doors. The Government recently published a consultation document, which is proposing a more transparent process, so hopefully it will be more open in the future.'

'How do big pharma companies select their Research and Development priorities against the 30,000 diseases that affect mankind and the 480 now being researched?'

'I can only speak about my own company, AstraZeneca. Our R&D into therapeutic areas represents about 80% of diseases in the developed world.'

'How true is it that diseases in the underdeveloped world have been neglected?'

'Many of the diseases in the under-developed world require simple old remedies. If we were to put all the necessary products on the docks in South Africa, they would probably sit there, or worse still, they would reappear in the western markets as parallel imports. Anything we do has to have Governments alongside so together we can do something that makes a difference.'

'What case is there for a more formal mechanism by which pharma and bioscience can cooperate in R&D to reduce the time it takes to get NCEs (New Chemical Entities) to the market?'

'Bioscience and pharma are working together now. So what is the 'USP' for both of these? For bioscience, it is usually fleetness of foot. For big pharma, it is the ability to both scale up from a good manufacturing practice perspective and turn the development machine to deliver viable programmes. There are many examples of where small companies tried to deliver viable programmes themselves. These companies know that they cannot market their products maximally. If they try to take products down the development path without the skills and resources, they usually don't have the ability to survive a major failure.'

'Recent reports in the press about large political donations to the Labour Party have been linked to the award of big NHS contracts, like the one given to Powderject. What has the ABPI done to put in place stringent guidelines to ensure that such things never happen again?'

'I don't really think it is up to the ABPI to become involved in how individuals and their companies spend their money. That is between them, their shareholders, the Government and their conscience.'

'Returning to my first question about priorities, what will be the yardstick by which you will judge your presidency, when you complete it in 2004?'

'On one level, I would want someone to say that "he leaves it better than he found it". I think that is what we should do in any job we tackle. I would then come back to relationships, that at the end of the two years we as an industry have found a way of improving our relationships with the caring professions.'

Reprinted extracts from full length interview 'Building the Future of UK Pharma' which was published in *Pharma Times* April 2002. Copyright John Lidstone.

These five case studies, although they concern contrasting people and events, embody important lessons for everyone who, from choice or necessity, has to face the press and be interviewed. They have been chosen because they illustrate graphically much of the advice, guidance and warnings contained in this chapter.

Dealing with the press: a checklist

(1) Is there a compelling reason for granting or seeking an interview with the media? If there is not, then do not give one.
(2) What are the objectives of the interview, both yours and those of the journalist?

(3) Never give a press interview without meticulous preparation.
(4) From the start to the finish of the interview, answer the questions as you planned to. Make only the statements that you intended to make and then *shut up*.
(5) Always answer the questions in as few words as possible so that editing is difficult, if not impossible.
(6) The longer the interview, the greater the danger that you will let your guard down and say something unplanned and damaging. If, for practical reasons, an interview has to extend over a period of more than two hours, arrange to have a break away from the journalist conducting the interview.

CHAPTER **10**

Preparing for a Television Interview

When you are approached by any of the media to comment or contribute to a programme, you have to decide whether to do it or not. Even if special circumstances leave you no option, *do not give an immediate answer*. Say that you will call back within thirty minutes and make sure that you do; journalists have a high regard for people who can be relied on to meet deadlines they have set and always do. Equally, they soon lose patience with those who don't. Use the time you have won to seek advice within your organization as to whether to do the programme and, if the answer is yes, then who ideally should do it. You may be working for a company with a clear policy on such matters laid down by the public relations department or press office. But this is still the exception rather than the rule.

Deciding to do the interview

Ask yourself and your colleagues:

- Will the organization benefit from taking part in the programme?
- If the answer is yes, are you the best or most appropriate person to do it?
- Are you available to do it, not just on the day, but for the preparation needed?
- Do you have the necessary knowledge and skill to do it?
- Does your company want you to do it?

Ask the media:

- What do you want from the programme?
- Why do they want me to do it?
- What role in the programme would I be expected to play?
- Will the programme take the form of an interview, or will I be a member of a panel or discussion group? Who will be the interviewer?
- What do I know about the interviewer or the members of the panel?
- When will it take place?
- Where will it take place?
- What is the name and status of the caller in the media company?
- What are the name and telephone number, fax number and e-mail address of the producer and director of the programme?
- How long will the programme last?
- Who decided to do this programme and why?

The answers to these questions should give you and your colleagues sufficient information to weigh up the advantages of doing the programme – and the advantages of not doing it. Remember that you are under no obligation to give an interview, but unless these are compelling reasons against it, be inclined to say yes. If your answer is going to be 'no', you should have a good reason that can be quoted to the media if need be.

Gathering information from the media

Armed with the background information you now have, you should return the journalist's call and find out a great deal more.

- Why are the media doing this article, broadcast, programme?
- Why have they contacted me?
- What questions will they ask? Do not be surprised not to receive a definite reply to this question. They are not simply being evasive. Often they prefer only to outline what the programme will cover in case providing a detailed list of questions leads you to give ready-made replies that will rob the programme of spontaneity and flow.
- Is there a hidden agenda for the programme? You might not get a direct answer to this question, but you can draw your own conclusions from vague or evasive replies. These you can share with colleagues to find out their assessment.
- In what context will my contribution be used?
- Will the programme be live or recorded?
- If recorded, will it be on film or videotape? (Film is rarely used except for magazine programmes.)
- Will film or props be used or required? They are an added complication, so beware. Thorough preparation and rehearsal are needed.
- Who else will be appearing on the programme?
- Why will the others be there? Am I going to be used as a scapegoat for whatever has gone wrong?
- How do these questions relate to the stated objective(s) of the programme?

Preparing yourself

Part I of this book offered a number of useful tips on how to prepare yourself. Here I want to emphasize a number of points that you should bear in mind:

(1) For a five-minute interview, set aside at least one hour for preparation. The importance of preparation cannot be stressed too greatly.
(2) Plan what you want to say, not what the media want you to say.
(3) People usually remember up to three points, so do not present more than three. If there is one point that is more important than any other, make this one the last and deliver it with great emphasis. Then, if the viewers or listeners only remember one, they are likely to remember this one.
(4) Likewise, never make more than three sub-points.

(5) Use brief, memorable anecdotes. Relate fact to a memorable story to aid the listener. 'Angina can be likened to pipes that fur up'.
(6) Use analogies, for example 'about the size of a football pitch' rather than 'one acre'. Visual pictures are easier for the audience to relate to and subsequently remember.
(7) Learn your brief and stick to it. Then you will always have something relevant to say and a framework to revert to if you become confused. Look back at the structure recommended for presenting your case in Part I.
(8) Never use jargon. The specialist language used between people in the same field of work may not be understood by anyone else. One of the mistakes often made in training pharmaceutical company sales staff is the failure to stress that they should refrain from medical terminology when talking to GPs and consultants. Similarly, avoid using those cursed abbreviations, initials or acronyms that have become part of a daily life. To most people not acquainted with them they are meaningless.

Anticipate the interviewer's approach

Many of the questions you are likely to be asked by an interviewer can be predicted:

- What did they tell you would be asked?
- Put yourself in the shoes of the interviewer. What questions would you ask yourself about the subject? (NB: Few journalists have any business knowledge or firsthand experience of it. Their background is usually the media and their career objectives are different from yours. Never assume that they are motivated by the same factors as you are.)
- The interviewer does *not* have the right to ask you questions outside the subject matter. If this happens, do not answer them. Ensure that the journalist returns to the subject and questions relevant to the programme and to your brief and why you are a part of it.
- Do not be defensive or apologetic. Use the journalist's questions to convey your message. Use their questions to your advantage. Mrs (now Baroness) Thatcher was skilled in responding to a question by saying, 'I'm glad you asked that question. It underlines the point I wish to make, which is . . .' Interviewer: 'I must stop you there, Mrs Thatcher'. 'No, you must not. Strong leadership . . .' Avoid defensive behaviour, such as clenched fists and tightly folded arms. They can give the interviewer, rightly or wrongly, the impression that you are holding back or have something to hide.

Dress and appearance

It is not unknown for those for whom appearing on a television programme is a new experience, or even when asked to take part in a radio interview, to buy a completely new wardrobe. Nothing could be more likely to make you feel uncomfortable and awkward, when you should be aiming to feel at ease.

Here are some guidelines on how to dress for such occasions:

(1) Avoid new or 'trendy' clothes. Wear non-contrasting items that will not distract the viewer's attention.
(2) Be your presentable self.

(3) Avoid narrow stripes or bold checks or very broad stripes, any of which could prove distracting.
(4) Do not wear jewellery that might reflect the strong lighting found in a TV studio. For a radio programme, do not wear multiple bracelets or anything else capable of making a noise that might be picked up by the microphone.
(5) Wear light pastel colours. They are much better than dark colours on television.
(6) If you wear spectacles, either keep them on or take them off and leave them off. There is nothing so distracting as someone who is always removing them while speaking. For the viewer this type of movement creates conflicting messages. Above all, never, never wear dark glasses. They are the surest way of indicating that you have something to hide!

Dealing with crisis

When you are interviewed, it is often a problem to say what you want in the limited time available. This is especially the case on television. There is no time for lengthy explanations when you are asked to give your views in thirty seconds as a news flash from an outside broadcast reporter. Your statements then must be clear and simple and convey a picture in a few words.

Yet such statements need to be the result of meticulous preparation. Nowhere is this skill more important than in a crisis. The checklists in Figures 10.1 and 10.2 are reproduced with the kind permission of the late Peter Tidman. They originally appeared in the MCB University Press 'Selling and Management Series', Volume 5 No. 2.

Creating a checklist is a useful way of preparing yourself to *face the press* at a time of crisis. The checklist in Figure 10.1 was produced for a well known oil company which was subjected to a catastrophe exercise by the Department of Energy in 1986. Some 67 issues were raised which the company's executives had to resolve in their statements. When the Department of Energy inspectors carried out the exercise they actually raised less than half the points for which statements had been developed. But a checklist like this allows you to be adequately prepared to face the interview.

Figure 10.2 shows a checklist of questions drawn up in advance of an interview by the management of a water company facing privatization. The quantity and quality of the water available was under scrutiny after a recent drought and the contamination of supplies in Cornwall and elsewhere. Moreover, the decision to privatize the water industry was hugely controversial. The company decided to prepare for everything the media might let fly at it. They brought in an outsider to help their executives brainstorm the questions the media might be expected to ask.

Similar models have proved of immense value to organizations involved in crises such as a take-over bid, the privatization of a car manufacturer, the Airtours disaster at Manchester Airport, public enquiries on Stansted Airport, Sunday opening of retail stores and the withdrawal of a pharmaceutical product following research that revealed serious side effects.

The pharmaceutical industry probably needs to be prepared more than any other for unforeseen crises. Misfortunes in one company can spill over and affect every other manufacturer of prescription medicines. The animal rights activities in relation to Huntingdon Life Sciences illustrate only too well the consequences for every other

Fire at night – shut down and get off the platform

1. What happened?
2. Is it contained?
3. Any casualties?
4. Head count (200).
5. Who is missing?
6. Evacuation by helicopter/boat.
7. Helicopter time to hospital/ashore.
8. Boat time to hospital/ashore.
9. Burns unit.
10. Emergency services ashore.
11. Spokesperson at emergency centre.
12. Radio/telex/tied lines to incident room.
13. Tied line/telex to Press Centre from incident room.
14. State of seas.
15. All lifeboats launched (radio in lifeboats).
16. Any visitors out there?
17. Can helicopters land on platform?
18. Pollution threat.
19. When will fire be contained?
20. What are you doing to find missing people?
21. What about NOK*?
22. Who is informing them (police)?
23. *Will* there be an explosion?
24. Danger to other platforms out there.
25. How long can men survive in water in survival suits?
26. Extra helicopters searching (Bristow).
27. RAF helicopters available?
28. NIMROD availability?
29. Help from other oil companies (Sector Club).
30. What went wrong?
31. Who is to blame?
32. Past incident record (Ensure you annotate all previous incidents).
33. Other major incidents elsewhere.
34. Medical facilities on 'stand-by boat'.
35. Is there a doctor on 'stand-by boat'?
36. Any nurses on 'stand-by boat'?
37. Any help from Frigg field (30 miles)?
38. Any other ships/trawlers in vicinity to help?
39. Any Royal Navy ships in vicinity?
40. Is this sabotage?
41. Any evidence available?
42. Previous threats.
43. What happens to NOK* if men lost?
44. Can wives (NOK* see their men?
45. Do you have welfare people to visit NOK*?
46. Do you have transport to help NOK* see their men?
47. Compensation for personal losses.
48. Health and safety checks.
49. When last done?
50. What did they say?
51. Was everything implemented?
52. Are the men fit for tasks/emergencies?
53. Who is now in charge?
54. Where is helicopter?
55. Production restart.
56. Effect on UK supplies?
57. Who will put out the fire?
58. Fire control.
59. Cigarettes, pipes, cigars.
60. Sleeping quarters.
61. Cookhouses.
62. Chip-pan fires.
63. Any drugs around?
64. Type of chemical extinguishers.
65. Security checks.
66. Terrorist or sabotage threat.
67. Structural integrity of platform.

* Next of kin.

Figure 10.1 Catastrophe at an oil platform: issues raised in an in-company exercise

pharmaceutical company. It is vital then to put the record straight about the safety of research, and other operations in your own organization. Later I will give some examples of issues with which the media can unexpectedly confront you.

The importance of giving good interviews to the media is paramount. Unfair though it may be, in the mind of the public you can be guilty by association. The media will pose the question of how your own company would tackle the same situation or problem. Anticipating issues which may arise and preparing suitable answers will help you to respond to any unwelcome surprises coming from an interviewer or any other protagonists in the studio.

Even then, you need to brainstorm every possible implication. For some years I have contributed to the MSc faculty of Pharmaceutical Medicine at a leading British university. A decision had been made to construct a purpose-built building for a psychiatric research unit

- What changes will occur on privatization?
- You will have the same people, technologies and procedures so how will you improve your operations?
- How will you inject a new motivation into people?
- Are you looking for high flyers to join you?
- Your government dowry is £150 million and your debts are £400 million. How will you clear those obligations?
- Will you manage your finances better in future?
- Will the same financial managers be in charge?
- Why should you do better this time?
- Will there be financial training?
- Why are you still discharging aluminium sulphate into your rivers, despite the experience in Cornwall?
- How do you deal with runoff of agrochemicals and pesticides?
- How do you deal with industrial pollution of the water supply?
- What happens to runoff from sludge spread on the fields?
- Why do you allow 40,000 tonnes of sludge a year to be dumped off-shore?
- Why have you banned canoeists on your rivers in favour of anglers?
- How are you handling discharge from fish hatcheries?
- What are you doing about discoloration of the water supply?
- How are you combating scum, bubbles and worms in water?
- Is fluoride safe?
- Why are you spending £30 million a year on television commercials?
- Are you co-operating with Friends of the Earth, the Pure Water Society and other groups?
- Is radiation in the water a threat to health?
- Why is the cost of water likely to soar?
- What about the 25 per cent of your customers who already find charges too high?
- Are you contracting out labour?
- What are your managers doing to relate to the local communities?
- What is your policy on water meters?
- How much will they cost?
- How long will they last?
- Why aren't you making meters yourself?
- What reserves do you have for product liability emergencies?
- Can anthrax infect the water supply?
- Why do we run out of water during the shortest dry spells?
- What have you got to show for all the money spent since the big drought of 1976?
- If you've spent thousands of millions of pounds making reservoirs safer, does that mean they weren't safe before?
- How can you justify pushing up charges just to pay dividends to shareholders?
- What about redundancies to reduce the wage bill and push up profit?
- Why are sewers suddenly collapsing all the time?
- Are there dangerous reptiles alive in the sewers?
- Can they come up through lavatories?
- What are you doing about the growing rat population in the sewers?
- Does our water supply carry Weil's disease from rat urine?
- Is there a danger of sewer explosions from methane?
- What about hydrogen sulphide?
- What is happening to industrial water rates?
- Are you going to charge for clean water going in and effluent going out?
- Why do you give discounts to well-off people who pay in advance?
- Is water contaminated with heavy metals from industry?
- What are you doing about lead pipes in old houses?
- Do washing machines have check valves to prevent soiled water getting into the water supply?
- Are we getting lead poisoning in areas where the water is acidic?

Figure 10.2 Water privatization: a case study

- Are manganese levels too high?
- What will happen to all the land you own when you are privatized?
- Are you going to build desalination plants?
- What will they cost?
- Who will pay for them?
- Are you going to diversify?
- Can AIDS from hospitals infect the water supply?
- What are you doing to prevent Legionnaires' disease?
- Is there Hepatitis B in the water?
- What about botulism?
- Is your water causing a kidney stone epidemic?
- Why is there such a high level of E-coli bacteria in the shellfish in your area?
- What about silage discharges from farms?
- Are you planning any more dams or reservoirs?
- Are your groundwater sources polluted?
- Do you have enough staff to monitor abstractions?
- Are you going to sell what was public land to developers?
- Are you going to restrict land access to ramblers?
- What is going to happen to your salary costs when you go private?
- Is the government turning a blind eye to water problems by agreeing to accept lower standards from privatised companies?
- What will Brussels say to violation of European standards?
- How are you making managers more commercially aware?

Figure 10.2 (continued)

in which the MSc faculty could also be housed. The night before the official opening of the new building, members of an animal rights movement, having seen the word 'Research' but not the rest of the centre's name, broke into the building and set it on fire in the belief that animals were going to be used for medical research purposes. Afterwards, this movement issued a statement apologizing for having chosen the wrong target – after millions of pounds of damage had been caused! Always imagine the worst possible case and prepare for it.

CHAPTER **11** *The Day of the Interview*

If it is your first experience of appearing on television, you are likely to feel both excited and nervous. Do not be tempted to go to a party the night before to take your mind off what's in store! You will need all your wits about you.

Getting to the television studio

Allow ample time for the journey so that you do not arrive at the studios late or in an agitated state. It is much more sensible to ask the studio to look after your travel arrangements in both directions – to and from the studio. This is especially advisable if the journey includes flying from one of the major airports. Such arrangements will not involve you or your company in any expense. It is in the interest of the programme organizers to get you there. They cannot risk your being late.

What to expect

A certain amount of makeup is usual, probably no more than a light powdering to reduce face shine and redness in your complexion. However, even this may not happen, as studios aim to show people looking more and more natural on television, sweat on the upper lip included!

If you are told that you cannot meet the interviewer until the actual interview, find out why not. If you are not satisfied with the explanations given, do not appear. There may be a quite simple reason such as that the person concerned has been held up or delayed on another assignment. If so, there should be no problem about telling you.

Go over the ground to be covered in the interview or the whole programme. Make sure that the interviewer has a written note of your name, correct business or other title and your position in your company or organization beforehand. Whether by accident or by design, interviewers have been known to introduce someone to a television audience by giving them a position higher in their company than the correct one. On that basis the interviewer has made unwarranted assumptions about the individual's responsibilities. If this happens, point this out to the interviewer immediately and make sure that your credentials are mentioned again, correctly.

If your name is liable to more than one pronunciation, tell the interviewer exactly how you wish to be introduced. I once had to interview a man whose name was spelled like the vegetable 'onions'. When I met him and welcomed him as 'Mr Onions', he immediately corrected me by saying 'My name is pronounced Oh-ny-ons'.

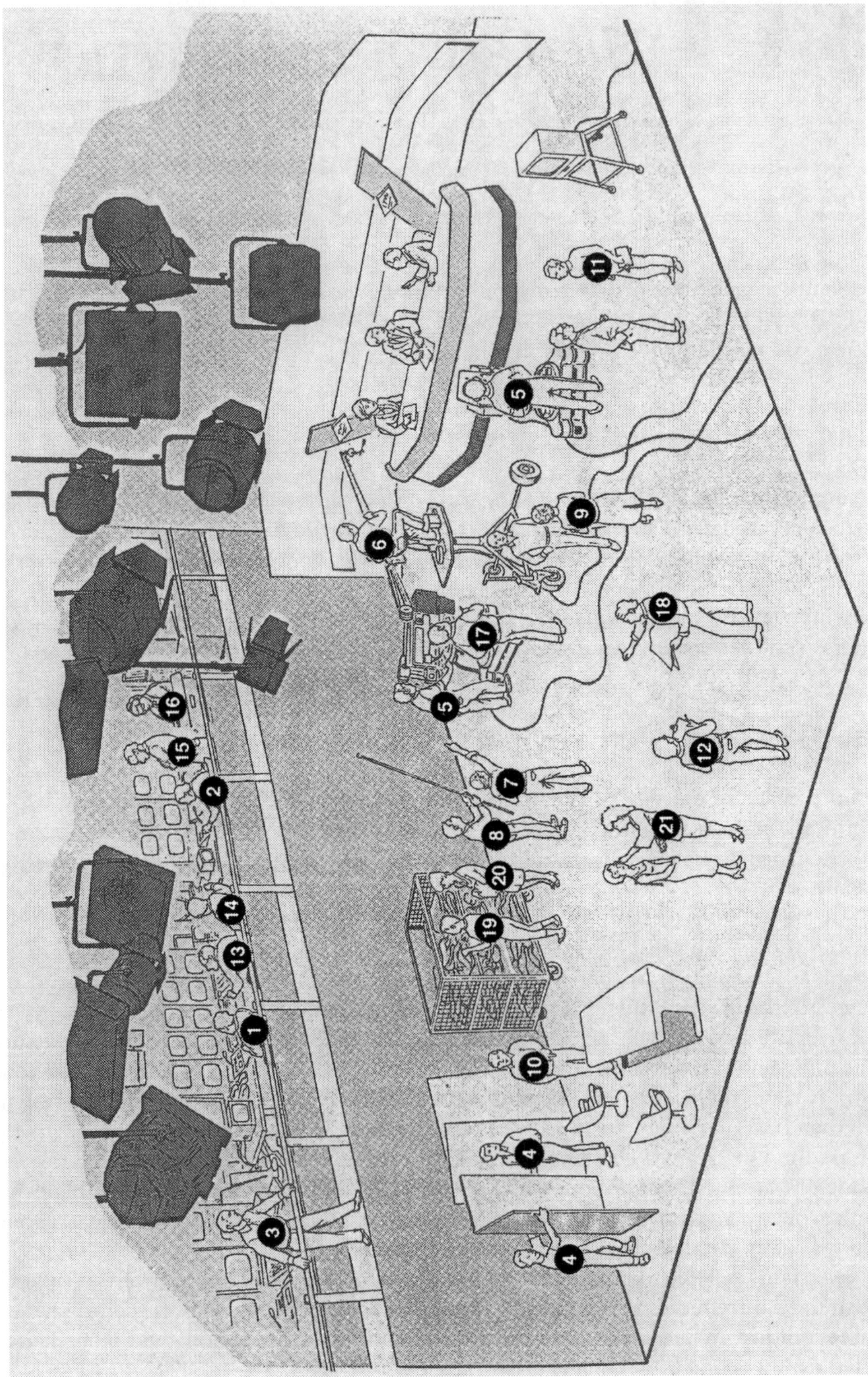

Figure 11.1 Studio layout

1 **The production assistant** provides organizational support to the director during the preparation of the programme and during studio operations, assisting with the timing of the programme ('calling the shots') and the subsequent editing and dubbing.
2 **The vision control engineer** is responsible for the electronic set-up of the cameras, working closely with the lighting director to ensure the best conditions for camera performance.
3 **The lighting console operator** is in charge of an electronic memory system which is pre-programmed to operate lamps at the right time during the recording.
4 **The scene hand** assembles and positions scenery on the studio floor, moving the different sets when needed.
5 **The camera operator** operates the camera itself on the director's instructions, following a pre-arranged script which gives the order of camera shots.
6 **The sound boom operator** positions the microphone suspended on the end of a boom to make the sound compatible with the picture, that is quiet for distant shots, louder for close-ups.
7 **The floor manager** makes sure that everything and everyone is in the right place at the right time on the studio floor, co-ordinating all activities and relaying the director's instructions to artists. The floor manager maintains contact with the control suite by short-wave radio talkback. Remember that the floor manager is in charge so put yourself in his or her capable hands, which are being paid to look after you. Let them do all the worrying instead of you. Ask to be taken through the whole procedure, for example, where to sit and who does what. If you want anything, just ask for it.
8 **The stage manager** ensures that everything on the set is in its designated place both before and during the recording, attending to the finer details.
9 **The floor assistant** aids the floor manager, ensuring that those participating in the programme are on the set when needed and giving them cues when it is their turn to appear or to speak.
10 **The producer** has overall responsibility for the programme, deciding the content, sometimes choosing the leading artists, and handling organization, administration and finance.
11 **The vision mixer** operates the vision control panel which cuts and fades pictures and produces special electronic effects, controlling pictures displayed on a bank of monitor screens. There is one for each camera being used, one for pre-recorded or filmed inserts and another for captions or still photographs.
12 **The operations supervisor** is responsible for the quality of both sound and vision, ensuring that all remote facilities such as video recorders or telecine machines are on hand. This may entail liaising with master control if a programme is going out live.
13 **The sound supervisor** balances one sound against another, ensuring that the quality of sound matches the picture by adjusting the tone and volume controls, and directs sound operations in the studio.
14 **The grams operator** plays in sound effects and music on cue, which are mixed with the speaker's dialogue from the studio floor. Many effects are added during later *dubbing* operations.
15 **The director**, who works in the studio on in the control suite, translates the script into action on the screen, directing actors and camera operators, then supervising videotape editing and sound dubbing after actual recording.
16 **The designer** researches the particular period of the programme to create the sets, draws the floor plan, constructs a working model and also decides the content of the set.
17 **The props hand** 'dresses' the set with furniture, pictures or curtains and supplies items such as books, food or telephones if they are required by the designer.
18 **The wardrobe dresser** 'dresses' the actors with costumes designed by costume designers to reflect the period accurately. Costumes are made by the costume department.
19 **The makeup artist** enhances the features of actors with makeup, ageing them or making them look younger, styling hair and applying effects such as artificial blood, wigs and scars.

Who does what

At the television studio there will be a variety of people engaged in making a programme, the editor, the producer, the director, the floor manager, the camera crew, researchers, secretaries and so on. The studio layout diagram in Figure 11.1, together with the accompanying notes, will give you some understanding of what all these people do.

Your notes

If you feel more secure with notes to hand, then just take a pad or cards with a few *large printed headings* on them. Never take whole sentences written down in longhand because when you glance down they will be difficult to read. Remember the warning in Part I against the danger of becoming a prisoner to your notes.

Make sure that the clipboard or notepad on which your notes are written can be held or balanced easily on your knees or a chair arm. Check such matters well beforehand.

Try to avoid having to read your notes all the time. It will lose you the interview with the viewers because you lose eye contact with them – and with it your credibility.

On questions of fact, a statistic or an important quotation, refer openly and confidently to your notes, reading out the specific figures or passage in measured words; for example, 'I want to mention some important figures relating to the research project we have been discussing, and I have them written down here'; then look down at your notes and read out what you want the audience and your interviewer to hear.

Voice test

Nothing is more distressing than being keyed up to answer the first question put to you by the interviewer only to find that you have lost your voice or, worse still, have a frog in your throat making you splutter out something incoherent. Having worked on film sets many times as a technical director, I repeat the advice offered in Part I. Before filming takes place, many directors ask the actors on set to have a good cough to clear their throats. It works, so do not hesitate to do the same, having checked beforehand that the red light for recording is not on.

You will be asked to take part in a short technical rehearsal so that technicians can carry out voice and camera tests. Do not say much and, above all, do not use this occasion to rehearse your opening words or the answer to a question. For a routine voice test, speak as normal. Do not shout, but keep the pitch firm and clear at the end of each sentence. Beware of *the English fade*. This is the tendency to tail off at the end of a sentence (see Figure 11.2). Be sure to pronounce the last words in each sentence clearly and at full pitch because they often convey the most important part of what you want to say.

Nerves

Speaking in public is a social skill. Since we all want to give a good account of ourselves, we care very much how successfully we exercise this skill. As a result, most people feel nervous at the start of an interview in which they are going to participate and also at the end. How can you minimize these two problems?

- By preparing answers to the types of opening questions likely to be put to you and then rehearsing by speaking through how you will deliver those answers.
- By preparing an opening statement that is short and clear and rehearsing it so that you are word perfect and do not have to refer to any notes to prompt you. If you do this, you

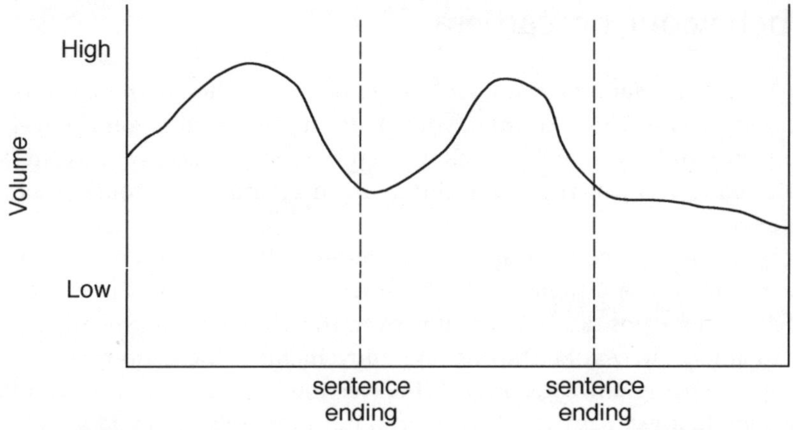

Figure 11.2 The English fade

will be looking at the interviewer and thus at the viewers and the result will be a picture of confidence.

- By preparing a short final statement which you have rehearsed so that if and when the interviewer turns to you and says, 'Please sum up your case in thirty seconds', you can respond clearly and confidently.

Verbal behaviour on camera

Speak at a slightly slower pace than in normal conversation. Remember that we speak at about 130–150 words per minute, so slow down to about 100 words a minute. Slow right down to about 60–90 words a minute when talking to an audience of mixed nationalities whose first language is not English. Remember that most of them will need to translate what you say into their own language before full understanding is achieved. And if what you say is being translated via earphones you will need to speak at a pace that allows the interpreters to translate all of what you say and not some of it.

Speak in short sentences. The shorter the better. If the programme in which you are participating is recorded, it is much more difficult to edit short sentences. Broadcasting is different from public speaking. It is not speech-making any more than it is normal conversation.

There is no time for lengthy explanations, which will simply invite the cut-off, 'Sorry, but we must stop there', leaving you with an unfinished and probably incomprehensible jumble of words as the viewers' final impression of you.

Television is an expensive medium. So go straight to the point as a newspaper headline does. On average, a television interview in a news programme lasts between two-and-a-half and three minutes, and your input could be less than that. On radio, a spot on a news programme like the BBC's *Today* averages the same or less. So there is no time for preliminaries, such as 'First I would like to thank you for the opportunity to speak to all your listeners' – you *are* speaking to them, so get on with it! Keep to the *spoken* English rather than that used in writing.

Physical behaviour on camera

If you have a 'good side' that makes you look better than the other, then let this be what the viewers and interviewer see. Sit forward in your chair. In this position you will feel alert and, what is equally important, you will look alert when the camera is on you. Avoid swivelling, or comfortable chairs into which you can slump down. Sit still, with your shoulders square to the interviewer.

Do not look at the camera. Look at the interviewer. Talk to the interviewer and you will be talking to the viewer or listener: it is the studio technicians' job to ensure that this happens. Their camera positioning, usually over the shoulder of the interviewer, will capture you in full view. Avoid shifting eye movements which, however unfairly, are regarded as a sign of being untrustworthy. Wherever you look or whomever you look at, do so for long enough to appear as though you mean it. Conversely, don't adopt a fixed glazed look like a rabbit trying to outstare a 15-ton lorry.

If you want to emphasize a point with your hands, do so provided the movement is not excessive. Your hands are a part of your personality so let them work for you.

Some teletips

Do not drink any alcohol before the programme even if it is offered to you in the studio hospitality suite. The slightest slurring of your voice will be very obvious on television.

Be on your guard from the moment you enter the studio until you leave it. When talking to the interviewer before the programme goes on air, do *not* say things like:

> 'This is not in my notes, but . . .'
> 'Although I was not going to say this in the interview, I feel . . .'
> 'This is off the record, but . . .'
> 'You won't say anything about . . . will you?'

Nothing is off the record to the ambitious interviewer who wants to make a name for himself or herself And if that has to be done at your expense they will do it.

If you used to watch the popular BBC programme *Yes, Prime Minister* you may recall the episode in which Civil Service mandarin Sir Humphrey Appleby is interviewed by Ludovic Kennedy. At the end of a dreadfully dull interview Sir Humphrey sighs with relief then, thinking that the interview is over, says what he really thinks. Out comes some scandalous gossip. Unfortunately he has failed to notice that the red light is still on: his every word is being recorded, ready for some future broadcast.

Always remember that at the studio you are on enemy territory. Everyone there is a potential news gatherer, aware that nervous people often speak unthinkingly. Be on your guard if you are left for some time alone in a reception room or hospitality suite. Such isolation, in such a charged environment, coupled with nervousness, can loosen the tongue and result in your giving unintentional ammunition to the interviewer.

If you have been inside a television or radio studio before, beware of passing yourself off as a 'professional' who knows the ropes. You may be left to your own devices and make yourself look foolish. Let the floor manager and studio staff show you to your seat and fix the microphone clips to your clothing. They have done this so many hundreds of time, so let them get on with it.

If you are unhappy about any aspect of the studio, such as the chair you are sitting in or the way some lights are shining on your face, or if you feel the need for a glass of water to be available during the interview, then say so.

Never lie on television. It is an unforgiving medium that will unmask a liar and hypocrite in ten seconds flat. Be yourself. However, if you have a quiet personality, then you may need to inject some additional enthusiasm so as to make an impact on the viewer.

CHAPTER 12 *Handling the Television Interview*

This chapter covers the actual process of being interviewed for television and explains how to make the most of your opportunities and avoid potential pitfalls. It also looks at the various different kinds of television interview.

Exploiting silence

As we noted in Chapter 9 it is the interviewer's responsibility to keep the interview going, not yours. If a silence ensues after one of your answers, then it is for him or her to cope with it. After you have finished answering a question, some interviewers will deliberately remain silent for longer than is necessary to induce unwary interviewees to implicate themselves. If you have no clear objective for the interview you will be tempted to fill the vacuum of silence created by the interviewer by saying things you did not plan to say.

Dealing with questions

Listen carefully to the whole of the interviewer's question without interrupting. Pause to give yourself time to compose your response, then answer the question. If an unexpected or difficult question is put to you do not be afraid to give yourself time before replying. Remember that pauses always seem longer to you than to your audience. However, in radio there is a difference and this point will be dealt with in the next chapter.

Hesitate or pause only if it is deliberate. Do not embroider your answers to questions. Know the limits of your own expertise and do not talk about anything about which you know little or nothing. Always be sure of your facts or have them to hand in your notes. One of the easiest ways to compromise yourself is when the interviewer allows you to go on talking.

SHUT UP

When asked a question, answer it. Say as much as you need to say, then shut up. Remember that the interviewer will be aiming to speak for only about ten per cent of the time, letting the interviewee speak for the other ninety per cent.

CORRECT MISUNDERSTANDINGS

If the interviewer rephrases your statements or answers, make sure that it is done accurately. If it is not, correct the misapprehension immediately and do not let the interviewer move on until you are satisfied. Do not be intimidated if the interviewer tries to change tack with a comment such as 'Time is running out so can we move on to another matter I want to ask you about.' Your reply should be polite but firm: 'No, you cannot, until I have cleared up the misleading impression viewers may have received from your interpretation of what I said.'

HANDLING INTERRUPTIONS

When answering questions put to you by interviewers, do not let them interrupt. *Stand your ground.* Alternatively, give way to the interviewer and then, when they have finished, continue with your answer as though they had not spoken.

KEEP CALM

Do not lose your temper or get involved in a verbal brawl. For the programme's director it makes great television, but not from your point of view, as it is happening at your expense.

REFUTE ERRORS

Refute any incorrect statements made by the interviewer or by anyone else taking part in the interview immediately and firmly. Do not even consider answering the next question until you have done so, for example, 'I cannot allow you to continue this discussion with me until I have dealt with the mistake you have just made. The research you referred to was not carried out by my company and you must withdraw what you said at once.'

At the start of an interview you can, if you wish, mention the name of the interviewer. But do not address him or her continually by name. Otherwise viewers and listeners will start counting the number of times you do so, rather than concentrate on what you are saying. Remember that the interviewer is there to get you talking, not to be the centre of attention or the purpose of the programme.

Developing a positive approach

There are a number of simple ways in which you can establish and maintain a positive approach.

DO NOT APOLOGIZE

Television interviews are seldom as bad as you imagine when heard or seen from the listeners' point of view. If you have made a mistake or you have to correct a fact, then do just that and move on without apology. This applies to live interviews.

STOP AND RETAKE

When an interview is being filmed or taped for a later broadcast and you make a mistake, do not hesitate to stop, explain why you have done so, and insist that your error should *not* be included in the transmission. The director then has two options, either to do the piece again or abandon the whole interview.

KEEP TO THE SUBJECT

Do not get sidetracked by the interviewer. And do not volunteer irrelevant information.

LOOK ALERT

Always look alert and interested, especially if you are taking part in a panel discussion or where there are others involved, even when you are not the subject of a question. Do not appear bored or stare at the ceiling. Turning to look at another speaker; sometimes nodding approval if something is said with which you agree, is a much better option. It also prevents the director, from the control monitors, recording any shots of you seen at a disadvantage.

BE PREPARED FOR SUPRISES

Try to prepare yourself for any surprises, for example a film clip shown about which you were given no advance warning, some statistics thrown at you or the unexpected introduction of another studio guest. If you feel confident about making a comment in a visibly relaxed way, you can put the interviewer in their place by saying, for example, 'Viewers should know that what has just happened was not planned to be any part of this programme. I wonder what the next trick will be?'

Verbal style

Never use jargon, abbreviations or colloquial English. As I mentioned in Part I, jargon arises from working in a specialized field of activity, such as medicine, the law, architecture, engineering or the armed forces. People operating in a specialized area develop the habit of speaking jargon to those who work with them and who will readily understand this technical shorthand. To the anyone outside that specialist circle their language is gibberish.

This can often be seen in pharmaceutical companies' medical sales forces, when for instance hospital representatives are calling on consultants to discuss a particular drug. On one occasion, I was observing how one of them was carrying out his work. During a meeting with a consultant psychiatrist, a hospital representative started using a number of medical terms that I did not understand. The consultant stopped him and said, 'Mr —, you are not a qualified doctor, but you are using terms which I am uncertain whether you know the meaning of or their relevance to what you are discussing with me. So will you please start again, but this time use language we can both understand.'

Many people find modish words and phrases like *cool, no way, at this point in time, at the end of the day,* meaningless or annoying. If they do not understand what you are saying then, as a communicator, you have failed.

Put yourself, always, in the shoes of the average listener or viewer and imagine how they think, feel and speak. Make what you say understandable to them and you will be communicating to everyone.

Mannerisms

We all develop mannerisms of one sort or another. And until we appear on a television programme or take part in a radio broadcast, we are never made aware of them unless we have frank friends or plain-speaking families.

- Avoid prefacing the beginnings of your answers to questions with 'fillers' like 'well', 'aha', 'um' and 'yes'. Unless what you say is compelling, your unseen audience will start counting the fillers instead of listening to the content.
- Do not end your answers with expressions such as, 'and so on', 'and so forth', 'etc, etc, etc.' They are irritating and pointless.
- Avoid continuous eye contact with the interviewer. It is uncomfortable for you both and it looks unnatural to the unseen viewers and listeners.
- Repetitive mannerisms, either verbal or visual, distract the listener or viewer from what you are saying. Instead of concentrating on what the interview is about, they start asking themselves 'What sort of person is this?'
- Visual distractions include lip-licking, wriggling in your chair, folding and unfolding your arms, removing and constantly replacing your spectacles, scratching you ear and running your finger across your nose. Don't fidget.
- Verbally, the constant use of particular phrases distracts, for example Sir David Frost's 'I mean', politicians' 'With respect' or 'The fact of the matter is'.

Some more teletips

EVERYTHING IS RECORDED

Assume that you can be seen and/or heard throughout the broadcast and immediately afterwards. Before you start relaxing make sure that the broadcast or recording in which you are taking part is *over* and the recording machine is off. For the same reason, keep still in your seat and avoid the temptation to lean forward to speak to the interviewer or anyone taking part in the programme. The final words you want recorded are not, 'Well, that was a stupid thing to say, wasn't it?'

Remember once again that the programme director in the control room has television monitors and can see and hear everything that is done and said. He or she will not hesitate to record an unguarded yawn, a raised eyebrow or a disgusted look to the ceiling which may be used to add colour or help make the programme more compelling. Your feelings will certainly not merit consideration compared with the chance to give an extra *frisson* to the discussion.

EYE CONTACT

Look at the interviewer or another participant when they are speaking to you. Beware in particular of looking away halfway through a question addressed to you. This will convey the impression, however unfairly, that you are not listening, even if you are. If you break your eye contact, look *down* rather than up to left or right and then look back at the interviewer. No one in normal life looks at another person for more than a few seconds at a time.

THE 'LAST WORD'

Do not try to score points. You have been asked to take part in a debate or programme because of your expertise, not as a comedian. Always remember that the professional interviewer controls the programme. He or she has the last word and is likely to win any verbal contest. Watch out for the interviewer who ends with some harmful comment like 'Listeners can draw only one conclusion from that answer. Goodnight.'

Know what the interviewer's deadline is for finishing the programme. Armed with this, you can sometimes lengthen your final reply so as to reduce their opportunity for making potentially damaging remarks.

NERVES

That tingle of apprehension, those butterflies in the stomach, are the signs that you care about what you are going to say and do. The opposite feeling, of over-confidence, can be lethal. Sometimes nerves come from fearing that you do not know the answer to a question. If you are asked about a particular matter and you do not know the answer, do not be afraid to admit your ignorance.

Self-assessment

Watch the programme in which you have participated as objectively and critically as you can and assess your strengths and weaknesses in the light of what you have read in this chapter. Check off the points on your brief for the programme. How many of them did you manage to impart as planned?

Different types of television interview

Not all television interviews follow the same format or the same procedures. As you would expect, what happens is governed by the purpose and nature of the programme involved. Let us look at those you are most likely to encounter.

1 NATIONAL NEWS PROGRAMMES

If you are asked to appear on a national news programme because of your involvement in a matter of current importance or because of your specialist expertise, your contribution will probably be a brief one. Judge the amount of time given by watching the early evening television news programmes on BBC or ITV. When the newsreader turns to ask an expert for a comment, the time allowed can usually be measured in seconds rather than minutes.

News items average one or two minutes. So every second you are given counts. After the newsreader or interviewer has introduced the item in which you will participate, there will probably be ten or fifteen seconds left for your comment or answer to a question. Therefore every word must count and none must be wasted. There is no time to say, for example, 'Thank you for asking me to give my opinion.' You do not need to do so. Your status has been established in the eyes of the viewer by virtue of having been invited to make your input.

2 LOCAL NEWS PROGRAMMES

In each television region the national news bulletin is followed by a regional programme made up of topical local items. Should you have an opportunity to take part in such a programme, you will usually have more time in which to air your views or to comment. In addition, the interviewers tend to be less aggressive than those on the national news networks.

Nevertheless, be factual, use short sentences and answer questions briefly and simply. A good performance can lead to further interviews – remember the journalist's 'little black book'.

3 DOCUMENTARY PROGRAMMES

Documentaries take the form of well-researched, detailed treatment of a specific subject. The programme maker will probably devote many days or weeks to background research. If you are asked to take part, it will be because of your specialist knowledge of the subject being examined. You need to have a clear, written brief on your contribution so that you know what is expected of you.

Often your contribution will consist of perhaps an hour-long interview. If the programme is recorded and the producer wants a number of views to be expressed, your interview may reach the screen as a total of five minutes, divided into three or four excerpts of 30–60 seconds from that one-hour interview.

4 CURRENT AFFAIRS PROGRAMMES

There are many kinds of current affairs programme, the best known being *Panorama* and *Newsnight*. If you are asked to take part in one, *be on your guard*. Such programmes tend to rely on dramatic revelations for their impact and appeal.

If the studio making the programme has an axe to grind, or a particular bias, ask yourself whether you should even be thinking of taking part. If you do so, insist on a live interview where there can be no editing of your contribution. Then prepare yourself – very thoroughly.

5 'DOWN THE LINE' PARTICIPATION

'Down the line' is covered fully in the next chapter in the context of radio interviews, but its television equivalent involves special features and problems that you need to be prepared for.

The television studio in which you will be installed is usually some distance away from the interviewer. In it there will be a television camera and sometimes, but not always, an operator. It is like one of those booths where you can have a passport photograph taken and developed. If no operator is present the camera will be operated by a remote-controlled joy-stick.

The interviewer may be miles away. You, in your booth, are connected to them by means of a British Telecom line or microdish. You cannot see the interviewer, but you can hear their remarks and questions, usually through an earplug or speaker.

This unnatural situation results in the sometimes startled, wooden expressions you see on the faces of people being interviewed from these 'down the line' studios.

Remember

(1) Look directly at the television camera lens.
(2) Keep still throughout the interview.
(3) Do not look at the television camera lens all the time because it becomes very tiring and difficult.
(4) When you look away from the lens, look down at your notes and then back to the camera lens. Never look upwards.
(5) Sit upright and sit forward in your seat so that you look alert.
(6) Sit square to the camera lens.
(7) The camera lens will highlight your appearance so, before the television goes live, ensure that
    - your hair is neat
    - your clothes are ironed with no creases
    - if you are wearing dark clothes, that any hairs or specks are removed as they will be magnified by the camera
    - if possible wear light-coloured clothing.

CHAPTER 13 *Preparing for and Handling a Radio Interview*

Although much of what I have already said about television interviewing applies to radio as well, there are a number of issues specific to radio, and these are dealt with in this chapter.

Differences between radio and television

On television, viewers can hear what you say, and they can also see you. You are communicating to them using two of your five senses. With radio, all that listeners can do is *hear* what you say. There is no picture to get in the way.

Radio is by and large a friendlier medium than television; the interviewer is not normally trying to best the interviewee. Radio is of course a much cheaper medium, so the opportunities are greater for extended interviews in which you can put your case fully.

There has been an explosive growth in local radio in the last few years. One disadvantage of this is that you may find yourself being interviewed by someone with little experience of the business world. On the other hand, there are more opportunities for you to develop links with your own local radio station. Indeed if you are in a specialist field of activity, you have a chance to become a regular source of comment.

Know your audience

When you are invited to take part in a radio programme, either for interview or perhaps to make a one-minute comment in a programme like *Today*, always ask yourself, 'Who listens to this programme?' The most effective contributions are made by those who are able to make themselves *personal to the listener*, who are felt to be 'speaking to me'.

AUDIENCE NEEDS

Beware of pigeon-holing your listeners into groups such as retailers, chemists, housewives, children, motorists, politicians, the young, the elderly, bankers, the disabled and so on unless you know that the programme is actually directed at one of these groups. If you do categorize in this way, then all those who do not perceive themselves to be in the group to which you appear to be speaking will switch off.

TALK ABOUT 'YOU'

On radio, always talk to the audience about 'you', because for the listener that means 'me'. Never use 'one', 'I', or 'we'. These terms all draw attention to yourself and away from the listener. Remember that an interesting person is one who speaks about other people; a bore only speaks about himself or herself.

When you are asked to take part in a radio programme or to make a comment, your decision will often be based on your desire

- to put forward a point of view
- to reinforce the idea that people should continue to pursue a certain line of behaviour or conform to a sensible regime, like a diet or disciplined exercises
- to suggest changing a harmful behaviour, as in, say, motoring or a dietary discipline. This means that you are trying to convince your listeners. Techniques for communicating your ideas were discussed fully in Part I. Here I want to emphasize three points that will help you win over your listeners:

  (1) Attention – Start by *talking* about the *listener* and you will gain the listener's attention.
  (2) Involvement – *Involve* the listener and you will gain the listener's interest.
  (3) Action – Aim to *finish with a powerful point* or *a request for action*, for example, 'This is what you should do if you want to make sure that your local hospital stays open.'

Preparing for a radio interview

Radio and television interviews require much the same kind of preparation, but there are some important differences to bear in mind.

SET OBJECTIVES

Have a clear objective:

- What do you want your listener to know?
- What do you want your listener to believe?
- If relevant, what do you want your listener to do?

Settle on two, never more than three, key points to communicate. If you choose three points and there is one above all else that you want to be remembered, put it last. Listeners tend to hear and remember the first and last things said to them. If your contribution comes towards the end of a programme and there is a danger of being cut off, forget the three points; just make sure that your listeners hear the one that matters most.

Use anecdotes, or short vivid stories, that illustrate success or failure. They are vital in conveying your message. Create simple, easily understood word pictures. For instance, if you were trying to describe an asthma attack, you might say: 'An asthma attack is rather like the constriction in the flow of water that would result from squeezing a hose pipe. The human equivalent is the windpipe.' Or, 'Gout is like the rusting up of the joints.' Or, trying to convey my childhood pleasure of buying fresh fish: 'When I was a small boy, living at Felpham in Sussex, we always knew that the mackerel brought round on the fish cart was fresh because it had sand on it from the nets being dragged up on to the beach from the fishing boat by old Boniface the fisherman and his two sons.'

Since radio is so much simpler than television, it is tempting to respond to an invitation to give your views on the spot to a caller from a radio station, especially if asked, 'Could you give your opinion on this issue now, over the phone, because we are going out on air later today?' As with television, always say, 'No, I cannot do it now, I'll call back in about thirty minutes.' Buy time to consult others as well as to think about what you will say.

REHEARSE

Rehearse what you are going to say and, just as importantly, how you are going to say it. There is no thinking time on radio. Long pauses are unacceptable, for a number of good reasons, not least because a long pause from the unseen listener's point of view might suggest that the programme has gone off the air!

Carry out a dummy run, using a colleague to listen to you. Record it on a tape recorder. Play it back and then both of you can analyze it. Find out how long the interview is to be and prepare for it working within the time limits. Never assume that you will be given more time. On the contrary, prepare on the basis that there will be less: if you are told you have eight minutes, aim to say what you want to say in seven.

Handling the radio interview

A radio studio is much smaller than its television counterpart, so try to sit with your back to the operations or recording room so that you are not distracted during the recording of the broadcast by all the activity taking place there.

VERBAL BEHAVIOUR

Avoid long pauses. Unlike television, there are no picture shots to take up the viewer's attention and interest while you collect your thoughts. As with television, you will be asked to do a voice test. Speak in a sustained volume throughout the radio interview. The modern microphone is so powerful that you have no need to raise your voice. Even so, do not move about in your chair or turn away from the microphone if a table-mounted or suspended one is used.

Face the microphone all the time. Speak clearly, maintain your pitch and remember not to drop your voice at the end of each sentence (remember that 'English fade'). Speak distinctly and not too fast. The listeners have no visual clues, so the whole message is received through hearing. Even though you will not be seen, sit forward in your chair throughout the interview as for television. It will keep you alert and bring alertness to what you say.

SINCERITY AND ENTHUSIASM

You have only one means of communicating on radio, through your voice. Through it you have to convey everything you think, feel and believe. Be enthusiastic. Enthusiasm always communicates itself to listeners – and if you are not enthusiastic, you should not be doing the interview or taking part in the programme at all.

Sound interested. Do not attempt to talk about anything you don't believe or don't know about. Remember that although you yourself may have heard your comments on a particular subject many times before, it will be new to the listener, so put energy into your delivery. Vary the tone and vary the length and pace of your words and sentences – they are the means by which you are going to create visions and word pictures to capture and retain the attention and interest of your listeners.

Types of radio interview

Many categories of programme are the same as for television, but there are factors applying to radio that should be borne in mind.

1 FACE-TO-FACE

You and the interviewer discuss an issue. Remember: avoid, if you can, sitting facing the recording room and the operators. Better always to have your back to it because of the distractions.

2 PANEL

In panel programmes a group of experts, or a group of people with conflicting views about a specific topic or issue, are brought together by the producer or radio presenter, who is hoping for a lively debate. Here you have to speak up and speak out when you want to make a point as there will be no director ensuring equal air time and opportunity to all the participants. It is worthwhile listening in to some examples of this type of radio programme beforehand to get a feel for how people talk, interrupt and make their views heard.

3 'DOWN THE LINE'

A 'down the line' programme involves being collected in a radio car, often somewhere far distant from the radio station, and asked to comment on, or respond to questions raised by, other sometimes equally disembodied and isolated programme participants or interviewers. You need to have a well-prepared brief to cling to so that you can express clearly and firmly the points you want to make. Sometimes you may have to interrupt someone to avoid being overlooked or even forgotten.

4 TELEPHONE

From the interviewee's point of view a telephone represents one of the most convenient ways of being interviewed. You are usually on your own territory, at your own desk or, at weekends, available at home. Again, you need to handle requests for telephone interviews so that you have ample time to prepare. Rather than answer there and then, tell the caller that you will ring back within five minutes to indicate whether or not you can do the interview. You can always say that you are in a noisy office and need to find a quieter one from which to speak. If the radio station presses you to speak to them immediately, be firm. Let them compromise instead of you. They may have deadlines to meet, but so have you.

Then use the time you have won to think about the subject the radio wants to discuss with you. If possible consult others in your organization, especially if issues could be raised about which you do not have up-to-date knowledge, for example, if you are the managing director and a reporter wants to talk about thefts of drugs made by breaking into medical representatives' locked car boots. You may wish to check the facts with your marketing director or national sales manager before allowing yourself to be questioned about such a sensitive subject. Alternatively, you may decide that one of these two executives should handle the interview.

Build goodwill with the journalist or interviewer who telephones you. Always promise to ring back with your answer rather than give a blunt refusal without thought. And then make sure that you do ring back. Journalists remember those who keep their promises – and those who don't.

Notes

Prepare yourself for the telephone interview by printing large headings on reminder cards which can be propped up and read without difficulty when you are speaking. The guidelines for such notes are the same as those set out for making a presentation in Part I. But for a telephone interview, you need to keep your notes as brief as possible and above all readable and easy to follow.

Avoid all distractions

Make sure that the office from which you are going to call back and speak to the radio interviewer is a quiet one and that you will not be interrupted by incoming telephone calls or other people. Tell your switchboard not to put any incoming calls through to you for at least 45 minutes. Ask reception and your secretary to do the same for anyone wanting to visit your office. Your choice of office should also take account of those on either side of it. Noises from another meeting, or a radio switched on to listen to your interview, can all be transmitted back down the line and ruin the radio reception.

Don't bellow down the telephone. Raise your voice to just above normal speaking level, slow down and pronounce each word clearly and crisply. These factors apply with even greater force when you are doing a telephoned interview for a foreign radio station. In such circumstances you need to check whether what you say is going to be translated as you speak. In such a case you need to slow down your delivery to about 90 words a minute.

Do not have a radio switched on in the same room from which you are making the telephone interview, because it will transmit its own signal down the telephone and so back over the air. The telephone interviewer will usually check this point with you.

5 OUTSIDE INTERVIEWS

Outside interviews can be conducted out in the open, at a luncheon table, in a car while driving or being driven, near the flight path at an airport, in a railway carriage . . . . Some producers like them because they add a touch of immediacy to the broadcast. Unless sound effects like traffic or aircraft flying overhead are desired, try to avoid distracting noises. You could end up with everything you planned to say being rendered inaudible; then what is the point of the interview?

Remember that all such interviews are likely to be recorded and edited before being broadcast. So if there are likely to be any comments of yours which could be cut or totally misunderstood, insist on a live studio interview.

Five tips for radio interviewees

(1) Write down the words you are going to start with and know them well enough so that you just have to glance at the headings you have made in advance.
(2) Prepare the structure of what you are going to say. Think yourself into the mind of the interviewer and imagine what questions you would ask if you were doing the interview.
(3) If necessary, use delaying tactics before answering a question, for example, by turning it back to the interviewer: 'Bearing in mind the subject we are here to discuss, why are you asking me what seems a totally irrelevant question?'
(4) If you find yourself in a muddle, don't hesitate to say, 'let me start again'. On pre-recorded programmes, if you make a mistake during the recording or get a fact wrong, do not be afraid to admit it. 'I'm sorry. I fluffed that; can we do that bit again please.'
(5) If you feel you have discussed a particular subject for long enough, do not be afraid to say, for example, 'Can we move on to another matter; I have said all I wish to about what we have just been discussing.'

CHAPTER **14**

Guidelines for Media Interviews

This chapter brings together information and guidance about media interviews from a variety of sources. It includes advice on what to do if you feel unhappy about the way you have been treated. The chapter ends with some final 'words of warning'.

Legal guidelines for radio and television interviews

Those responsible for radio and television programmes work within guidelines laid down by the British Broadcasting Corporation (BBC) and the Independent Broadcasting Authority (IBA). Three important features of these guidelines are:

(1) Whether the interview is recorded or live, interviewees should be made aware of the format, subject matter and purpose of the programme, as well as the way in which their contribution will be used. They should also be told the identity and intended role of any other proposed participants in the programme.
(2) Interviewees should be told if an edited version of their interview will be shorter than the original and the programme makers should take care that the shortened version does not misrepresent the interviewee's contribution.
(3) The context in which extracts from a recorded interview are used is also covered. An interview should not be edited so that, by juxtaposition, a contributor is associated with a line of argument which he or she would probably not accept and has had no opportunity to comment on. Neither should separately recorded interviews be edited together so as to give the impression that the contributors are in actual conversation with each other.

Sir Robin Day's Code for television interviewers

The late Sir Robin Day, one of television's most experienced interviewers, gave me permission to reproduce his Code here. It was first published in 1961 and proudly reprinted in his memoirs, *The Grand Inquisitor*, in 1989.

Readers should bear in mind the year this Code was compiled. As Sir Robin put it, 'My use of the pronoun "he" should not be taken as indicating disregard for women interviewers. In 1961, one was forgiven for adopting the old legal, if sexist, maxim that 'masculine embraces feminine'. That apart, the reader may judge whether it has stood the test of time, and whether on television, it is more honoured in the breach than the observance.'

Here is the Code:

(1) The television interviewer must do his duty as a journalist, probing for facts and opinions.
(2) He should set his own prejudices aside and put questions which reflect various opinions, disregarding probable accusations of bias.
(3) He should not allow himself to be overawed in the presence of a powerful person.
(4) He should not compromise the honesty of the interview by omitting awkward topics or by rigging questions in advance.
(5) He should resist any inclination in those employing him to soften or rig an interview so as to secure a prestige appearance, or to please authority; if after making his protest the interviewer feels he cannot honestly accept the arrangements, he should withdraw.
(6) He should not submit his questions in advance, but it is reasonable to state the main areas of questioning. If he submits specific questions beforehand, he is powerless to put any supplementary questions which may be vitally needed to clarify or challenge an answer.
(7) He should give fair opportunity to answer questions, subject to the time limits imposed by television.
(8) He should never take advantage of his professional experience to trap or embarrass someone unused to television appearances.
(9) He should press his questions firmly and persistently, but not tediously, offensively or merely in order to sound tough.
(10) He should remember that a television interviewer is not employed as a debater, prosecutor, inquisitor, psychiatrist or third-degree expert, but as a journalist seeking information on behalf of the viewer.

When I talked to two well-known interviewers for the BBC Today programme about Sir Robin's Code, their remarks were interesting. First, John Humphrys, giving a lecture at a City Livery in 1997 entitled: 'I may not agree with what you say but I will . . .', said, in response to a question from me, that he was not sure that Robin Day always followed his own precepts. But then he added, 'Of course he and I have always tried to get straight answers to relatively simple questions. To achieve this sometimes requires aggressive questioning.' He then cited the former Cabinet minister whose response to a particular line of questioning was, 'The last reporter who tried to take the piss out of me got a terrible shock.' Humphrys' final comment was, 'Interviewers should always use simple language'.

Secondly, Sue MacGregor, in a review of one of my books on how to deal with the press, said this of Sir Robin, 'I am sure Robin stuck to his code.'

Fourteen ways to deal with the press

An article on managing, in *Fortune* magazine in June 1989 said that a war on the press, tempting as it may sometimes be, cannot be won. Then it listed the following guidelines which are reprinted by kind permission of *Fortune* Magazine, © The Time Inc. Magazine Company 1989. All rights reserved.

(1) Make the chief executive officer responsible for press relations.
This means the officer must often speak for the corporation both routinely and in times of crisis, and delegate enough authority to make the public relations spokesman a credible source.

(2) Face the facts.
If you screw up, admit it candidly. Avoid hedging or excuses. Apologize, promise not to do it again, and explain how you are going to make it right.

(3) Consider the public interest in every operating decision.
Your reputation depends far more on what you do than on what you say. Act accordingly. Try giving your senior public relations expert a seat at the table when decisions are made.

(4) Be a source before you are a subject.
The time to make friends with reporters is long before trouble hits. Get to know the people who cover your company, educate them, help them with their stories and give them reason to respect you. Determine which journalists deserve your respect and trust.

(5) If you want your views represented, you have to talk.
Reporters are paid to get stories, whether you help or not. When you clam up, they must depend on other sources – often people like that marketing vice-president you fired last month.

(6) Respond fast.
You cannot influence a story once the deadline has passed. Nor will you appear credible if you seem to be stalling. In a crisis, figure you have a day to get your story out.

(7) Cage your lawyers.
They will always tell you to keep your mouth shut. However, in many crisis situations your potential legal liability may be trivial compared with the risk of alienating your customers, employees, or regulators.

(8) Tell the truth – or nothing.
Nobody likes a liar.

(9) Don't expect to bat 1.000.
Public relations is a game of averages, so be content if you win most of the time. Even the most flattering story will likely have a zinger or two, and even the best companies get creamed now and then.

(10) Don't take it personally.
The reporter is neither your enemy nor your friend; he or she is an intermediary between you and the people you need to reach. Forget about your ego; nobody cares about it but you.

(11) Control what you can.
Release the bad news before some reporter digs it up. Use your selective availability to reporters as a tool. Set ground rules every time you talk. If the public isn't buying your message, change it.

(12) Know who you are dealing with.
The press is not monolithic. Television is different from print, magazines are different from newspapers and the *Austin Statesman* is different from the *Wall Street Journal*. Within a news organization there will be a normal mix of individuals, some honourable and competent, some not. Do your homework on journalists before you

talk to them, reviewing their past work and talking to other executives they have covered.

(13) Avoid television unless you feel free to speak candidly.
Even then, learn to present your views in ten-second sound bites that are the building blocks of television stories. Use simple, declarative sentences and ignore subtleties. Whenever possible favour live television shows over those that can edit your remarks.

(14) Be human.
The public will usually be more sympathetic to a person than to a corporation. If you can do it without lying or making a fool of yourself, reveal yourself as a person with feelings. Your mistakes will as likely be forgiven as criticized. Insist on being judged on a human scale, with normal human fallibility taken into account. Remember that people love to root for the underdogs.

The same article contained a short glossary for talking to reporters. Here it is.

Talking to reporters: a glossary of terms

Negotiate the ground rules with reporters before you volunteer information. Knowing these definitions will help you.

(1) Off the record.
Material may not be published or broadcast, period.
Do not go off the record casually or with anyone you do not have reason to trust.

(2) Not for attribution.
Information may be published, but without revealing the identity of the source.
Always specify whether that applies to your company as well as to yourself. Nail down the attribution the reporter will use – 'A member of Acme Corp's two-man executive committee' versus 'an industry expert' – before you open your mouth.

(3) Background.
Usually means material not for attribution. Do not take this for granted. Discuss it with the reporter.

(4) Deep background.
Usually means off the record.

(5) Just between us.
And other ambiguous phrases mean little to reporters. Do not use them.

(6) Check it with me before you use it.
Means just what it says. Specify whether the restriction applies to quotations as well as facts. When the reporter checks back, you have the right to correct errors and misunderstandings, but not to withdraw statements you now regret.

(7) Read it to me before you use it.
Gives you no right even to correct errors. All you get is advance warning of what the reporter will use.

(8) No.
Means that you have decided not to answer a reporter's question. Used judiciously, this can be a life saver.

The article ends with the following wise words.

> The bottom line, which is not exactly news, is that dealing with the press means dealing with perception. That may be uncomfortable for executives used to communicating hard facts. Certainly trying to help convey perceptions through the filters of reporters, editors and producers is risky. But it is less risky, on balance, than not trying.

Complaints

If you are unhappy with the way you have been treated by a newspaper or a programme, discuss it with your management colleagues and, if there is one, the head of your public relations, or public relations agency if you retain one. If you are still not satisfied there are procedures in place for pursuing a formal complaint.

THE PRESS

If your complaint concerns an item in a newspaper involving inaccuracy, intrusion, harassment or discrimination, write to the editor of the publication in question. If you receive no satisfaction you can take up the matter with the Press Complaints Commission, an independent organization charged with enforcing an editorial Code of Practice for the press.

RADIO AND TELEVISION

Take your complaint in the first instance to the Director General of the BBC or the Independent Broadcasting Authority and then, if necessary, to the Broadcasting Complaints Commission. Viewers or listeners who have a complaint which they wish the Commission to consider should write to the Secretary giving the title of the relevant programme and the date and channel on which it was broadcast. They should explain in what way they regard the programme as unjust or unfair, or in what way they believe that their privacy was unwarrantably infringed. If they were not a participant in the programme, they should also explain their interest.

When the Commission have considered and adjudicated on the complaint, copies of their adjudication and a summary of it are sent at the same time to the complainant and the broadcasters involved. It is the Commission's normal practice, whether or not the complaint has been upheld, to direct the broadcasters to broadcast the summary and to publish it in *Radio Times* or *TV Times* as appropriate. The summary of the Commission's adjudication is usually broadcast on the same channel as, and at a similar time to, the programme which was the subject of the complaint. This is the only sanction available to the Commission; they cannot require the broadcasters to apologize to the complainant, to broadcast a correction or to provide a financial remedy.

If all else fails and you decide to take legal action, do so quickly, using a lawyer with experience in media litigation who will not have to undergo a long learning process at your expense.

Final checklist

(1) Establish in the first place what your involvement is going to be in any interview or programme.
(2) Brief yourself, and any others in your organization whom you are consulting, on your involvement.
(3) Before it takes place, establish what type of programme you have been invited to contribute to or participate in.
(4) Find out as much about it as possible.
(5) Obtain confirmation that none of your comments will be used in a repeat programme or out of context without your prior permission.
(6) Know your rights as defined by the BBC and IBA guidelines.
(7) If you or any part of your organization is to be filmed, make sure that the film crew is accompanied throughout any visit(s) by a senior member of your company. This person must be briefed about the purpose of the filming, the objectives, any no-go areas, or questions that might be asked by the film crew.
(8) During the filming, make sure that you are clear and precise in dealing with questions. Be as short in your replies as possible. Beware of how your answers could be edited.

Training in Media Relations

Many readers, having completed Part II of this book, will be asking, 'How can I develop the knowledge and skills required to face and handle the media effectively?' Whether you are the managing director or chief executive of a small company, or one of several senior directors or executives of a large pharmaceutical firm or a multinational or a public organization, this appendix offers you guidance on the training options available.

External courses

For the senior executive of a small company it is usually too expensive or impractical to have a training programme specially designed for yourself alone. However, there are a growing number of external courses devoted to helping senior personnel to handle the media.

Here are some guidelines to help you identify the course that best meets your training needs.

(1) Define your training objectives in terms of:
   (a) What you need to know and understand about the media.
   (b) What you must be able to do: for example, brief journalists at company annual general meetings, issue and be prepared to comment on press releases to the media, handle interviews with the media in person, on radio and television, on the telephone, down-the-line ...
   (c) What standards of presentation and interviewing skills you must attain to be able to undertake media interviews.

This analysis will help you determine exactly what you must be capable of at the end of the training programme.

(2) Send for information about these courses on how to handle the media. Check the following with the organizers:
   (a) Who actually conducts the programme? What are their qualifications? They should include first-hand experience as a journalist, perhaps someone also with some organizational or business experience like your own, and above all they must have been trained to train.
   (b) Which companies and organizations have used the course and do so regularly?
   (c) How many attend each course? Since your objective is to develop yourself, you want to be sure that you will receive individual tuition. Ideally, there should not be more than six delegates on such a programme, preferably with two tutors and the

use of professional radio and television interviewers to put you through simulated press, radio and television interviews.

(d) Ask the course organizers for the names and telephone numbers of at least three people in positions similar to your own who have attended the course which you decide best suits your needs and to whom you can speak about the programme, its quality and effectiveness.

Individual tuition

As an alternative to attending an external training programme, you could arrange for private tuition from an acknowledged expert just before you are going to take part in a press, radio or television interview. I have undertaken such tuition for politicians, Cabinet ministers and heads of organizations, especially when they are doing it for the first time and in strange conditions, such as through an interpreter.

The benefits of private tuition are:

- It can be related to the individual's specific interviewing needs and topics such as with the president of the Wellcome Japanese subsidiary described in Chapter 9. One of my training assignments involved following a Cabinet minister around as he spoke to different audiences all over the country, taping his speeches and then sitting down with him over a weekend, analyzing his strengths and weaknesses and agreeing what he had to do to overcome some of his presentational weaknesses.
- The tutor can help you to script some of specifically-tailored statements and play devil's advocate on key topics you are likely to encounter in facing the media.
- The tutor can coach you on an individual basis. The costs may be high, but they should be outweighed by the benefits, not least because the coaching would be built around the specific interview situations you will be facing immediately afterwards.

An in-house programme

Many companies, public associations and organizations employ, or have as members, a large enough number of senior people to justify designing their own in-house training programme. Such a programme can be organized either by the central management development or human resource manager or by commissioning a specialist in media training.

How to structure a company programme

Here are some guidelines on how a 'How to handle the media' workshop might be constructed. The term workshop is used deliberately because during it learning experiences take place; knowledge, systems, techniques and skills would be developed to the point of practical application in situations likely to be faced by the delegates at the completion of the workshop.

WORKSHOP OBJECTIVES

By the end of the workshop, delegates should be able to:

- set clear objectives for media interviews and at the same time understand those of the media
- control media meetings and interviews so that the televised, broadcast or published feature(s) do not contain anything said in an unguarded moment and regretted later
- handle sensitive issues so that correct facts and impressions as conveyed to the public
- prepare, structure and conduct all types of media interview.

CONTENT

Based on these objectives, the material covered in the workshop would include:

- handling the media: an opportunity for delegates to discuss their individual perceptions and experiences, if any, in dealing with the media and why they have arrived at such perceptions
- press, radio and television interviewing: the case for and against doing them; the key features of press and television interviews; how to prepare and structure such interviews; preparing structures for such interviews; television studio and outside broadcasts; what to expect; who does what; your appearance and manner
- handling the interview; the rules that apply; key points; self-assessment
- radio and telephone interviews; the differences between television and press interviews; how to prepare yourself; types of interview – face-to-face, panel, down-the-line, telephone on tape, outside
- guidelines governing radio and television interviews
- the press: articles for publication; the process and the pitfalls; editing and ability to comment; using a tape recorder for interviewing.

METHOD

The workshop should be designed to allow the maximum amount of time for discussion and questions arising from each subject after it has been presented and developed by the tutor.

The bulk of the workshop, at least seventy per cent, should be allocated to developing media interviewing skills, ideally using all the technical equipment now deployed by radio, television and the press. This will include creating a television studio with two or three cameras so that delegates can get a lifelike experience of studio conditions, additional rooms with television cameras and recording equipment for one-to-one interviews and telephones wired into a tape recorder so that telephone interviews can be recorded and analyzed.

If the tutors are not highly skilled in all types of television, radio and press interviews, then you need to import one or, preferably two, professional interviewers. Some of the interviewers used by the BBC and independent television news programmes are freelances and, provided they are given sufficient advance notice, might be willing to undertake workshop interviewing of delegates. Their fees, depending on their reputation, range from £250 to £1,500 per day.

ROLE-PLAYING

The most effective way to develop skill in being interviewed is through repeated practice. To achieve this, each delegate should have to handle a variety of media interviews during the workshop, ideally dealing with subject matters related to his or her field of activity.

These role-plays should include one-to-one interviews with a professional journalist, one-to-one telephone interviews by a journalist seeking material for an article with a tight copy deadline, emergency telephone interviews about a crisis at, say, a chemical plant or a business scandal such as questions in the press about future but so far unearned fees from providing interim surgery management teams having been wrongly included in the company's annual accounts.

As so many pharmaceutical companies operate worldwide, these crisis telephone interview situations should require delegates to be woken in the early hours of the morning or summoned from a meal. This will test how effectively they can manage unprepared for requests for comment.

In role-plays, the studio atmosphere, with a number of television cameras and powerful lighting, will conjure up the actual experience people go through when being interrogated on television.

All the interviews should be recorded and played back to the delegates so that prior perceptions can be discussed. Their individual strengths and weaknesses can be analyzed and suggestions made as to how performance can be improved. Some of the one-to-one interviews should be analyzed on an individual basis and comment on a delegate's mannerisms and behaviour can be offered in a constructive and candid spirit.

LENGTH AND NUMBERS ON WORKSHOP

To cover the material and effectively develop the media interviewing skills of delegates, numbers on the type of workshop described should be limited to six, based on two tutors and one or two professional interviewers (if the tutor(s) need support in this area).

The workshop should ideally be held on a residential basis over a period of three days. This will allow time for all the activities I have listed and, in particular, for individual practice interviews, playback and evaluation.

CASE STUDY MATERIAL

Any case studies used on the workshop should reflect the circumstances for which delegates are being trained. Either delegates should bring with them details of actual media situations they are about to face after the programme so that the workshop literally helps them to fashion their material in advance. Or the tutors must prepare case studies that simulate real life.

Here are three case studies related to the pharmaceutical industry.

Case study 1

You are the head of your company's medical information department. The company has moved on to a new site where security is all-important. You have been asked by the managing director to brief all staff on how to handle a possible raid on the research buildings by an animal rights movement.

Your task

(1) Prepare and write down a checklist of 25 questions so that you can brief your colleagues on what they might be subjected to by the media in such a situation.
(2) Use the questions you have prepared for a press interview you will carry out.

Case study 2

You are the medical director of a pharmaceutical company which routinely uses toxic materials.

Problem The local press is reporting that some highly dangerous toxic waste has leaked from your production plant into a stretch of the Medway river where children often swim during their spring and summer holidays. Some have been reported as having developed severe skin disorders.

Your task Your secretary has received a telephone call from the editor of the *Kent Messenger*. He is planning to write an article on this matter and wants you to return his call so that he can discuss with you: 'how a company like yours, concerned with the environment and public health, can allow toxic material to continue to be used in a plant which is so near a major river – especially when you must know what happened to Sandoz at Basel, when a leak contaminated the Rhine'

Identify at least 15 questions which you believe the editor of the *Kent Messenger* is likely to ask you, so that you can be ready for an interview at which these and possibly other questions will be put to you.

Case study 3

You are the marketing director of a leading pharmaceutical company and interested in responding to an invitation from the *Financial Times* which is planning the content of next year's Review of the Pharmaceutical Industry. Your company has a major product in the treatment of osteoporosis and the FT would like to interview you about this medical condition, particularly in elderly women.

Your opportunity The FT journalist, though unwilling to give you the questions to which she wants you to respond, has indicated the approach the article will be adopting. She is interviewing a number of pharmaceutical companies with products in this therapeutic area, and wants to know, among other things: how many women over 55 years of age suffer from osteoporosis?; what are the costs to the NHS of orthopaedic operations as a result of broken bones, shoulders, hips, arms, wrists and so on arising from this condition?; what actions your company is taking, if any, to tackle this huge problem in terms of preventive measures/medicine?

Your task To identify the questions the FT journalist is likely to ask you and to work out how you will answer them. This is a big therapeutic market and, with other 30 other pharmaceutical companies in it, you will have to make some telling and competitive points. Apart from consulting your company colleagues, you will need to be armed with information about what other companies are doing.

After the workshop

The tape-recorded interviews should be saved and given to the individual delegates at the end of the workshop. This will enable them to compare their actual performances when they face the cameras and interviewers with their workshop practice runs. They will be able to see what progress or mistakes they have made and note what they need to correct in future interviews.

Media perception questionnaire

At the beginning of Part II you were asked to complete a media perception questionnaire as frankly and objectively as possible. Now, using your photocopy of the original blank questionnaire, please fill it in once again. This time you will do so with the benefit of having read the material in Part II on how to handle the media, whether press, radio or television.

The reason for repeating this process is of course to enable you to compare your perceptions of the media as a result of a better knowledge and understanding of what journalists do. Have they changed? If so, for the better or for the worse – and why?

References and Recommended Reading

Clare, J. (2001) *John Clare's Guide to Media Handling,* Aldershot: Gower.

Day, R. (1961) *Television: A Personal Report,* London: Hutchinson.

Day, R. (1989) *Grand Inquisitor,* London: Weidenfield and Nicholson.

Lidstone, J. (1985) *Making Effective Presentations,* Aldershot: Gower.

Lidstone, J. (1992) *Face the Press,* London: Nicholas Brealey.

Mackay, I. (1984) *A Guide to Listening,* London: British Association for Commercial and Industrial Education.

Margach, J. (1978) *The Abuse of Power,* London: W. H. Allen.

Pease, A (1981) *Body Language,* London: Sheldon Press.

Tidman P. and Lloyd Salter, H. (1992) *Tidman's Media Interview Technique,* Maidenhead: McGraw-Hill.

Index